ENCYCLOPAEDIA OF HORTICULTURAL CROPS

Volume-3

CULTIVATION OF HERBS AND FLOWERS

S. K. GUPTA

RAJAT PUBLICATIONS
NEW DELHI - 110 002 (INDIA)

RAJAT PUBLICATIONS
4740/23, Ansari Road, Daryaganj
New Delhi- 110 002 (India)
Phones : +91-11-23267924, +91-9811484973
E-mail: rajat_publication@yahoo.com

Encyclopaedia of Horticultural Crops

Edition 2025

ISBN 978-81-7880-319-7 (Set)

[The responsibility for facts stated opinion expressed or conclusions reached and plagiarism, if any, in this volume is entirely that of the Editor. The publisher bears no responsibility for them whatsoever.]

PRINTED IN INDIA

Published by Mrs. Seema Wasan for Rajat Publications, New Delhi and Printed at H.S. Offset Press, Delhi.

Preface

Horticulture refers to the area of agricultural science that deals with garden plant cultivation, the science and out of growing fruits, vegetables, ornamental trees bushes and flowers, on a commercial scale, in nurseries, orchards and gardens. Along with forestry and agronomy, horticulture constitutes the three broad branches of agricultural science and practice. As importance and relevance of aggressive cannot be ignored, so can that of horticulture be also not ignored. In fact, with the increase in demand for newer and better varieties of horticultural products, the significance of horticulture has increased all the more.

This encyclopedia of horticultural crops is a three volume set which brings a compendious compilation of information on the various forms science and practice of cultivating horticultural crops. Through these three works, foundation has been laid by the encyclopedia for a discussion of the principles of horticultural science, what it comprises of, its history and its present, and what the future holds for this field.

The encyclopedia commences with the book organic vegetable production a book which seeks to discuss the cultivation of organic vegetables in agro-science. As organic farming and its products have entered the market in a significant manner, it becomes imperative that a manual seek the principles of this practice and therein lies the purpose of this book. Cultivation of Herbs and Flowers as the title suggests, is a work bringing forth a comprehensive detailing of cultivating herbs and flower which after vegetables, consultant a major part of the demand for horticultural products, both for purposes in practical applications (as in medicinal), and for ornamentation. The concluding work in this series Is the book Horticultural Crop Production. As many horticultural crops today particularly vegetables—are often grown as field crops and considered agricultural commodities, it has become increasingly evident that horticulture products are much more importance than as mere garden products. This book seeks a study of horticultural science as a whole, with special emphasis on its products as "Crops".

All these three books aim at providing a reliable encylopaedia on horticultural productions; and provide material for through study of the science itself. It is hoped that the comprehensive and informative content of this series foster an enriching reading and learning experience for readers.

S.K. Gupta

Contents

1

Introduction to Herb Cultivation

The flavor and the aroma of a given dish, be it a famous French or Italian cuisine, or, for that matter, a less famous oriental dish, depend very much on flavorings and seasonings derived from herbs and spices. Both herbs and spices can give the simplest of foods a variety of flavors such that these foods can be served time and again in so many ways that one never gets tired of them. The story of spices as well as that of related plants, aromatic plants, has already been published. The present publication concentrates on 'herbs' in Asia.

Asian peoples, particularly Southeast and South Asian, are very fond of herbs. The Asian traditions of using herbs are based on centuries of trial and error. The accepted recipes were then passed on from generation to generation, many of which have been modified and improved by later generations.

Unfortunately, herbs are, in act, very bitter. Examples: sweet marjoram, sweet basil, celery. unlike western peoples who often described the use of herbs in their dishes in written records, very little has been recorded by Asian peoples in spi'e of the fact that oriental foods, in general, are much more 'spicy' (as the result of added flavors from herbs) than western foods. Among them are

Indian, Indonesian and Thai foods, all of which are quite popular, even among European and American, due to the presence of herbs which are added to the dish as a garnish. Herbs have a dual role to play in the kitchen; they are decorative and useful - attractive to look at, and invaluable for flavoring.

In addition to using as flavoring for food and beverages, herbs are also useful to mankind in many other ways, namely:

— *As medicine*: Most herbs can be used to cure human ailments. In fact, the term 'herb' is often understood by most laymen as 'medicinal plants', while 'herbalist' is equivalent to traditional doctor who treats his patient with medicinal plants based on various systems of traditional medicine.

— *As cosmetics*: The use of herbs as beauty aids can be traced back to the earliest civilisations. The ancient Egyptian were coloring their hair with a mixture of henna and indigo, and scenting it with marjoram 5,000 years ago. Oriental peoples also used various kinds of herbs to beautify their bodies from head to toe.

— *As dyes*: Both edible and non-edible dyes may be obtained from herbs. These are used in food coloring, hair coloring, and textile coloring.

— *As air freshener*: Sweet scented herbs, either grown in the vicinity of the house, or hang up bunches or branches, provide safe, natural air freshener.

— *As disinfectants*: Essential oils and some other chemicals present in some herbs have both deodorising and disinfecting prop. rties.

— *As insect repellents:* Many herbs, such as mint, pennyroyal, basil, holy basil, citronella, lemongrass, have insect-repelling properties.

— *As botanical pesticides*: These are organic compounds derived from some plants, including herbs such as citronella, holy basil, lemongrass, used in controlling pests of crops, animals, and humans.

— *As decorative materials:* Many herbs are decorative in the sense of Tussie-Mussies, an old English tradition of using herbs for decorative purposes, the most traditional and attractive of all herb decorations, adornments, and devices.

— *As herbal drinks and teas*: Chrysanthemum, mint, basil, chamomile, etc. are frequently used in herbal drinks and teas. They are not only refreshing, but many also claim to possess therapeutic property; hot basil tea, for example, is helpful in alleviating indigestion and stomach cramp.

— *As pot pourri*: Dry herbs such as thyme, sage, rosemary, marjoram, together with dried fragrant flowers (e.g. rose, lavender, jasmine) and spices (e.g. cloves, cinnamon, nutmeg, coriander seeds) are mixed to give sweet scents to the air in a room.

One disadvantage of herbs over spices is that the former are derived from leafy parts of plants which, in normal conditions, must be used in a fresh state. Whereas spices are nonleafy parts, easily dried and processed and may be kept for longer period. Thus, herbs need to be cultivated by the users; they are rarely processed and sold in the market.

The gradual movement of people from the rural areas to industrialised areas meant a decline in cultivation and fresh harvest of herbs. This has stimulated the need for factory-processed herbs, which, in turn, initiated large-scale cultivation of herbs, accompanied by industrial processing into dried, ground herbs and readymade herb ingredients for instant cooking. Advance in food technology has made new food products like frozen food, canned food, instant-powdered food, dried-packed food, etc. quite popular, all of which require flavoring ingredients such as herbs to improve their taste. Recent Popularity of Herbs

Several factors have contributed to the recent popularity of herbs. These are:

Increasing international travel and immigrants: Convenience and rapid means of transportation have brought tourists as well as immigrants from one land to another. With them, access to different styles of cooking and flavoring offered by these immigrants from one culture, has re-awaken popular interest in herbs. The diversity and increased availability of herbs have encouraged many people to introduce dishes incorporating different flavors into their culinary repertoire.

Back-to-Nature Campaign: The growing dissatisfaction with artificial flavorings, seasonings, and colorings has motivated many people to turn to natural flavors, etc. of herbs. Health Conscious: The use of certain food ingredients, like salt and sugar, to add taste to food has been known to have certain health hazards. Reduction in the amount of salt and sugar can be achieved through the use of certain herbs.

Herbs in Asia

Asia is a region well known for its rich biological diversity. This includes diversity of plant crops of all kinds, due to the existence of five Centers of Diversity of cultivated plants, namely,

(1) the Indo-Malayan Center,

(2) the Chinese Center,

(3) the Indian Center,

(4) the Central Asia Center, and

(5) the Afghan Center.

Asia is also a region of ancient civilisations, and people had inhabited this part of the globe for more than five thousand years. They had made use of indigenous plants as food, clothing, shelter and medicine, the four necessities of life, since the dawn of history.

Asian foods are spicy because they are flavored with indigenous spices and herbs. Much have been written about spices of Asia, but very little about their sister crops, simply because herbs are special group of plants which must be consumed fresh, thus little known outside the region. In contrast, spices had been brought into Europe and other parts of the world by the Arab traders thousand of years ago.

It is fair to say that both spices and herbs had made Asian foods spicy. Since most literature do not distinguish spices and herbs, and the only term used has been 'spices', while the term 'herbs' has been hidden and forgotten. In the West, 'herbs' have been well known and recorded in words of written languages, and thus became a separate category of flavoring/ seasoning ingredients of foods. Since there are numerous species of herbs used in Asian countries, many of which are unknown to the rest of the world, or even to other countries within Asia, it is impossible to list all of them for lack of references.

Cultivation of Herbs

Characteristics

Although considered high-value crops, herbs in Asia are not as systematically cultivated as those in other regions.

The characteristics of herb cultivation in Asia are as follows:

Subsistence, mixed cropping systems: Most herb crops in Asia are cultivated by small holders who grow several kinds of crops including herbs in a mixed cropping pattern. This is quite adaptive to the existing conditions where a herb industry is not yet established. However, there are also a few large plantations of monocropped herbs like lemon grass and basil.

Labor intensive: Herbs are a special kind of crops which require intensive cultivation, thus very labor-intensive at all stages, starting from land preparation, seeding, transplanting of seedlings, to caring of plants throughout their life cycle. They also require quite heavy fertiliser application to obtain luxuriant growth in order to have a pleasant appearance to the consumers. In the case where synthetic chemicals are not allowed to be applied to the crops, even more intensive care has to be provided. When it comes to harvesting, much more labor is needed than ordinary crops. The same is true for post-harvest operation. Thus, herb cultivation probably is one of the most labor intensive ones. Women have played a major role at all stages of herb cultivation.

Use of primitive cultivars: In most countries in Asia where herbs are grown by small holders, primitive cultivars are used which result in poor yield and quality. In places where large-scale cultivation is taking place, improved cultivars are used.

Improper post-harvest treatments: Small holders naturally do not have knowledge of post-harvest treatments. This results in poor quality produce. The treatment is improved as the cultivation moves to a larger scale.

Genetic Improvement of Herbs

As a crop, herbs receive little attention from Asian breeders. The only country where genetic improvement is conducted is India.

However, only few varieties of herbs have been released for commercial cultivation in India. As for other countries, practically no varieties have been developed. Varieties of herbs grown in Asia are mostly materials introduced from other countries, the rest is obtained from native materials. Thus, there is a need to start a breeding programme for genetic improvement of herbs for commercial cultivation.

The following paragraphs summarise the approaches to obtain improved cultivars of herbs:

Introduction from other countries: This has already been practiced in several countries. It is the easiest and least time-and-effortconsuming. The only limitation is adaptability of the introduced varieties to local conditions.

Selection from already existing variants: Variation which occurs spontaneously in any population of herbs can be selected for future uses, either immediately or after hybridisation.

Traditional breeding: Using conventional crossing techniques, breeders can combine desirable characters from the two parents. Objectives of herb breeding are similar to those of other economic crops. However, as quality and quantity of active compounds responsible for flavor and fragrance of a particular herb play the most significant role, the major objective of herb breeding is high quality and quantity of the constituents.

Biotechnological methods: Advances in biotechnological research have facilitated de novo genetic improvement of herbs through:

(a) increasing genetic variability of the germplasm,

(b) culturing and selecting desirable genotypes,

(c) rescuing embryo of selected genotypes,

(d) rapid multiplication of clones of selected genotypes,

(e) transferring genes from distant parents, and

(f) specific gene transfer.

Factors Affecting Productivity

Herbs are special crops that require heightened attention from growers in order to obtain high productivity of quality products. Several factors affect their productivity, namely:

Soil structure and fertility: Herbs require well-drained soils of rather high fertility. Loamy soils with high organic content are ideal for growing herbs.

Sunlight: Although some herb species can survive under partially shaded condition, most herbs prefer full sunlight which is needed not only for maximal photosynthesis of the crops, but also to synthesize essential oils, the major constituent of herbs which are causal for flavor and fragrance. Both intensity and duration (day length) are of considerable importance for the synthesis of essential oils. Many species require long-day sunshine for optimum production of these constituents.

Temperature: Although many herbs can be grown in cool climates of the temperate zone, the ones that are grown commercially in Asia require warmer climate.

Pest control: While herbs are commonly believed to repel insects and diseases because of specific compounds found naturally in these plants, the accuracy of this belief is not entirely known nor complete. While the essential oils of basil and holy basil, for example, have been reported to have anti-microbial activity under controlled conditions with specific pathogens, many insects and diseases are found on such plants in the field. Hence, they should be checked regularly for insect and disease incidence. Most effective in the control of insects and diseases is the use of resistant varieties and integrated pest management.

Weeds are a serious problem for herb cultivation because they interfere with herbal growth. Moreover, weeds cause great nuisance when it comes to separating them out after harvest. The herbs will be 'off-flavored' due to the presence of weeds, whether consumed no matter fresh or processed.

Several brands of weed killers, both pre- and post-emergent, may be used to control weeds. Among the recommended weed killers is terbacil (Sinbar). If properly used, this broad-spectrum chemical gives season-long control of most weed species. However, because of residual effects, terbacil often poses a problem with other crops if used in a rotation programme.

Post-harvest Handling of Herbs

Herbs are very delicate, perishable crops which require special post-harvest handling to obtain high quality of fresh herbs, or even in processing. Even when excellent quality is achieved through proper harvesting and packaging, quality is seldom maintained throughout the marketing process. A number of factors need to be considered in optimising post-harvest of herbs. They include proper temperature, humidity, handling techniques, atmospheric condition and control of pathogens.

Temperature: Temperature is the single most important factor in maintaining quality of herbs after harvest. A 10°C decrease in product temperature may increase storage life by as much as 3 to 4 times. The quality of most herbs is best maintained at temperatures close to 0°C. For best results, herbs should be harvested in the early morning to avoid loss of moisture due to field heat, otherwise a pre-cooling operation is needed. This is normally done by forced air-cooling. If herbs are to be stored for a certain period, they should be placed at temperatures above 0°C to avoid freezing damage consequent to their high water content.

Humidity: Transpiration of fresh herbs results in early wilting and consequent loss of quality. To avoid transpiration, herbs should be packed in polyethylene bags and placed in cold rooms or air-conditioned transport vehicles where relative humidity is maintained at a high level.

Handling techniques: Most fresh herbs are susceptible to damage during post-harvest handling. Discoloration and physical injuries (e.g. bruise, breaks, etc.) are encountered if they are not handled with care. Microbial infection is likely to follow after bruise or injury.

Atmospheric condition: The occurrence of ethylene released from fresh herbs during storage may have severe effects on some herbs. These include injury symptoms, chlorosis, leaf fall and downward bending of the petioles. However, they will not occur if the room is well ventilated.

Control of pathogens: Herbs packaged in polyethylene bags are likely to be infected with pathogens. Control measures include pre-harvest and handling hygiene and avoidance of injury.

MAJOR HERBS IN ASIA

The lack of production statistics on herbs in most countries makes it very difficult to determine how large the numbers of herbs grown in Asia. Based on personal judgement and communications with Country Representatives of ANMAP, the author selects 17 species of herbs as major ones commercially grown in Asia.

Basil (Ocimum basilicum)

Basil is one of the most common herbs whose principal use is culinary. It is of Asian origin but also cultivated in southern, central and eastern Europe, North Africa, and also in the USA, particularly California. The flavor of fresh basil is sweet, fragrant and spicy. It is one herb that cannot be dried successfully since after drying, the taste is quite unlike that of the fresh herb. However, it can be quick-frozen although it loses its color unless blanched first. It is commonly used in several Oriental dishes.

It is an annual bushy plant, 30-60 cm. high with branching stems that bear tender, light green, oval leaves with slightly toothed margin. Clusters of small, white or pale purple, two-lipped flowers appear towards the tops of the stems. It can be grown from seeds or cuttings. It is normally grown in the backyard gardens but commercial production for fresh consumption does occur. Processing of dried leaves is also observed. The form of basil used to flavor food is different from the one that is used as fragrance that is commonly known as sweet basil. It is also treated as an aromatic plant.

Celery (Apium graveolens)

Celery has a pungent smell and a bitter taste. Both the leaf and the seed are used in flavoring foods. It is a biennial plant with a bulbous fleshy root and coarse, ridged and branched stems, 30-60 cm. The leaves vary in color from yellow green to dark green, and are borne opposite, with dentate margin. Flowers are borne on sparse umbels, greenish-white in color. Seeds are small, oval-shaped.

Coriander (Coriandrum sativum)

Coriander has been used as a flavoring and a medicine for over 3000 years. It has always been a favorite culinary herb in Asia and was introduced to Europe by the Romans. The green leaves have a fragrant, strongly aromatic taste that lends a characteristic flavor to soups, salads, curries and other Oriental dishes. They are extensively used in Southeast Asian and Chinese cooking to flavor vegetable dishes, with fish and meat, and as a garnish. The plant is also harvested for its seeds and used as a spice. In Thailand, the fleshy roots are also used as herbs in several Thai preparations.

Coriander is a hardy annual plant, 40-60 cm. tall with slender, branched stem bearing flat, parsley-like lower leaves and feathery, thread-like upper ones. The leaves are aromatic when crushed. Flowers are born in flat umbels, white to pale mauve. Seeds are round, and drop as soon as they ripen. Coriander is grown from seed. It needs a dry, light soil and a sunny, sheltered position.

Curry Leaf (Murraya koenigii)

The curry plant is a native of South Asia. It is principally grown in southern India for its aromatic leaf which is an essential ingredient and the predominant aroma and flavor in Madras curry powder. Many southern Indian dishes are flavored with its leaves and it is also popular in Southeast Asia and Africa.

It is a small shrub whose leaves resemble those of bay leaves (*Laurus nobilis*), one of the most common herbs used extensively throughout the world to flavor meat and fish dishes. In India, they are generally used fresh. In

Europe and the US, dried curry leaves are usually available from shops specialising in Indian and Chinese food. Fresh leaves can be dried in the oven or quick-frozen.

Dill (Anethum graveolens)

Dill has a faintly sweet taste. It is normally used in salads, and in particularly associated with fish dishes. It is very common in Vietnamese dishes. It provides flavors to other vegetables like peas and potatoes.

It is an annual plant growing to a height of 50 to 75 cm. and looks much like fennel, particularly its inflorescence, which is an umbel. It is grown from seeds on raised beds at a distance of 15 cm. between plants in a row 30 cm. wide. It requires full sun for optimum growth and flavor leaves are harvested when flowers just start to develop. For spice, its seeds should be left to mature on the plant until dry.

Holy Basil (Ocimum tenuiflorum)

A plant closely related to basil, holy basil is used as a herb, and medicinal plant. Except in Thailand where it is commercially cultivated as herb for Thai cooking, holy basil is only grown in home gardens as a medicinal plant. It is considered a sacred plant in India where it is used in religious ceremonies, hence the name 'holy' or 'sacred' basil.

Holy basil is an annual, bushy plant, 30-60 cm high, with branching stems that bear tender leaves with toothed margin. Two main color variations of the leaves are found, light green, and purplish green. The latter has stronger flavor. The inflorescence, which also exhibits the same color variation as the leaves, consists of small, two-lipped flowers, born at the top of the stems. It can be grown from seeds or cutting. Cultivation practices are similar to basil.

Its twigs can be harvested only one month after planting, and two weeks after every harvest during the growing season. The whole twigs are best used fresh for cooking or they can be dried and ground for storage.

Lemongrass (Cymbopogon citratus and C. flexuosus)

Lemongrass is a perennial herb, densely tufted, with dense clusters of linear leaves. All parts of lemongrass are lemon-flavored and the main ingredient (78%) is citral. The part used in cooking is the basal part of the leaf sheath that is a false stem of the plant. It can be used fresh or dried. The leaves can be used for essential oil extraction by steam distillation.

Lemongrass rarely flowers in cultivation. Propagation is by the use of the lowest part of the false stem (about 2-3 cm) attached to the underground stem. Once the plant is established, tillering will take place within two months, and more tillers will be borne. They are ready to be harvested any time after that.

Marjoram (Origanum marjorana)

Marjoram or sweet marjoram is an annual herb of the genus *Origanum*, many of whose species are aromatic plants. Many other species of *Origanum* having aromatic compounds used in flavoring or seasoning are also known as marjoram. These include *O. onites* (pot marjoram), *O. heracleoticum* (winter marjoram), and *O. dictamnus* (Dittany of Crete). Sweet marjoram grows well in a sunny site. It can be grown from seeds, either through direct seeding or through transplanting seedlings that are grown in the nursery rows or boxes. It grows to a height of 25 cm. The stems are tough and woody. The leaves are oval, dark grayish green. The flowers are very small, borne in clusters around the stem.

Mint (Mentha spp.)

Mint is probably the most common aromatic plant since there are at least 40 species of the genus Mentha, many of which possess several varieties of different morphology and aroma in their leaves. Species used as herbs include common spearmint (*M. spicata*), apple mint (*M. rotundifolia*), pennyroyal (*M. pulegium*), and more common in Asian countries than in other parts of the world, kitchen mint (*M. cordifolia*). Mint may be used in both sweet and savory dishes. Oriental people have spicy dishes sprinkled with kitchen mint.

Kitchen mint is propagated by cutting, stolon (underground stem) or layering. The latter is a natural phenomenon without human assistance. If the stems are covered with soil, roots will be formed and the rooted stems can be detached for planting. Commercial growing prefers to use stolons since they are hardier in transport and planting.

Oregano (Origanum vulgare)

A very close relative of marjoram, oregano is a hardy annual that grows to a height of about 20cm, with woody stems and dark green leaves, 2 cm long. The flowers are small and white in color. They are borne on long spikes.

Oregano thrives well on a well-drained soil in full sun but can also be grown indoors provided it is placed on the windowsill. Propagation is by seeds. The fresh leaves are used as salads and in various dishes. Dried leaves are also used.

Parsley (Petroselinum crispum)

Parsley is probably the best known herb because it is so widely used as a garnish (which is often thrown away by people who think that it is used only for decorating the dishes!), since it has beautiful, yellowish green, curled leaves. It has a unique flavor and can be used as a condiment or a persillade.

Parsley is a hardy biennial that flowers the year after it is sown. Seeds quickly lose their viability and should be sown as soon as possible after their harvest. Germination unusually takes a long period, to appear up to 8 weeks, but do much more quickly if the soil is warm.

Rosemary (Rosmarinus officinalis)

A native of the Mediterranean and widely cultivated in temperate climates as a garden plant, rosemary is a very aromatic evergreen shrub whose height may reach 1.5 m. Branches are covered with soft hairs when young and later become woody with grayishbrown scale bark. Leaves are long and narrow with a dark green upper surface and a pale gray, hairy underside. When rubbed, a strong camphor-like aroma is released. Flowers are pale blue in color, with two lips, borne in clusters on upper leaf axils of the branch.

Rosemary is propagated from seeds, but commercial varieties are normally propagated from cuttings. It prefers a sunny position in well-drained soil. It is a popular kitchen herb. It has strong pungent flavor.

Sage (Salvia officinalis)

Sage is an evergreen shrub, 50-90 cm tall. The young stems are covered with white hairs but become woody at the base with age. The leaves are thick, grayish green, oblong with round apex. Inflorescence is borne at tip of branches, consisting of tubular, two-lipped, violet-blue flowers. It is most adaptive plant in terms of soil condition. With the exception of very acid soil, it can grow in almost any type of garden soil. It is normally used as stuffing or forcemeat. It can be used on grilled pork chops.

Several varieties have been developed. Propagation can be made by seeds, cuttings or by layering. Sage prefers a well-drained soil and a sunny position. Being a perennial crop, sage can survive year after year by developing woody stems at base.

Sage is strongly aromatic and has a powerful flavor, especially when dried. It is one of the commonest kitchen herbs in European cooking.

Savory (Satureja hortensis)

There are two kinds of savory, summer (*Satureja hortensis*), and winter (*S. montana*). The first, which can be grown in Asia, is an annual herb that is self-sown. It has a strong, hot and slightly bitter flavor.

Summer savoy is propagated by seeds. Seedlings should be thinned to about 15 cm apart. It needs strong sunlight and good drainage. It grows to a height of 45 cm. The leaves are dark green, long and narrow (like soft pine needles). The flowers are small, pale mauve.

Shallot (Allium ascalonicum)

Shallot is closely related to onion (*Allium cepa*) and garlic (*Allium sativum*). It is one of the most common herbs used throughout the world. Most Oriental dishes are flavored with its leaves that give a fine tang to flavors, nuttier than an onion and distinct from garlic.

Shallot is very easy to grow; propagating by the use of small bulbs (also incorrectly called 'seeds') taken from the cluster of mature bulks (which are also used as spice). They are normally grown on raised beds, several clusters of young bulbs producing clusters of tube-long leaves that should be harvested when still green. Harvesting is done by pulling the whole plants including their young bulbs.

As propagating material, wait until the leaves fade completely, then pull the whole plants up and leave them to dry in the air until quite dry, then store the bulbs in cool dry place.

Tarragon (Artemesia dracunculus)

Native to southern Europe, tarragon is now cultivated commercially in Europe, Asia and the USA. It is a bushy, aromatic perennial, 60-90 cm high. The slender stems bear smooth, dark, shinny leaves; linear in shape, and widely spaced along the stem. Flowers are globular, grayish-green or greenish white, and woolly. Propagation is by seeds or cuttings.

Thyme (Thymus vulgaris)

Thyme is one of the most important herbs, used as a real all-rounder, in stuffing for meats and fish, in stews, casseroles, poultry, forcemeats, marinades and soups. It is particularly good on certain vegetables, onions, etc. Other species of *Thymus* are also used as herbs, such as lemon thyme (*Thymus citriodorus*).

Thyme is grown from seed on raised beds. It can also be propagated by means of layer. This is done by spreading the branches out from the center and covering their stems with soil, leaving just the leaf tips exposed. Roots will be formed within ten days and the rooted branches can be detached for transplanting.

Form of Herbs

The flavor and aroma of herbs are due to the presence of essential oils that are complex organic compounds produced by the plants in their glands located in their leaves and flowers. They are volatile compounds evaporating with heat. They also evaporate during storage, although quite slowly, especially if kept in cool place. Different kinds of herbs behave differently in their full release of flavors and aromas, depending on their condition and inherent characteristics. This difference is referred to as the aggregate form of herb that can be divided as follows:

Fresh Herbs

These are herbs that are best used in their fresh state since their flavors are easily lost soon after picking from plants. It is best to pick these herbs when they are needed for use. If bought herbs are used, they should be kept as fresh as possible. The best method is to store them in the refrigerator in sealed plastic bags or a box so that the leaves or young flowers do not dry out.

Dried Herbs

Herbs dried before use are most suitable for industrial processing and use, especially in food industry. They have a more pronounced flavor than in the fresh state. However, not all herbs can be dried; some are simply unsuitable for drying. Examples of such herbs are dill, Spray Drying: This is done through the use of modern equipment, the same method used in making spray-dried coffee.

The herbs are boiled and concentrated, and finally placed in the spray-dry apparatus. The product is in a powder, easily dissolved in hot water. parley, chives, fennel. The flavor of certain herbs, notably basil, is much impaired by the drying process. Some dried herbs also lose both scent and taste, becoming hay-like when kept for longer than 3-6 months. Commercially available dried herbs are either ground (powder) or chopped into small pieces. They are packed into small bottles equipped with numerous holes for sprinkling.

Freeze-dried Herbs

There are herbs that can be freeze-dried. Although considerably more expensive than conventionally dried herbs, they retain both flavor and color much better, and resemble fresh herbs since they have absorbed some of the cooking liquid.

Types of Herb Products

Unlike spices, which, by their inherent characteristics, favor processing resulting in numerous types of products, herbs are normally consumed fresh. However, a few herb products are available in the markets, mostly derived from dried herbs. These products include:

Ground Herbs

Once the herbs are dried, they can be ground into various-sized products, from coarse to fine and powder forms depending on the nature of herbs and the purpose. They are then packed in a small bottle, whose inner cap is perforated by holes that allow the ground spices to come through by shaking.

Herb Teas

Herbs are processed into a product ready to be used to make tea. There are two methods to make herb teas:

Increasing demand: Industrialisation and the resulting urbanisation make it impossible for urban people to grow their own herbs while demand is increasing due to rises in living standard and increases in population.

Tea Bag: This is made by packing small amount of ground herbs in special paper bags. Most is made of only one kind of herb although some may be blended. These bags can be made into a drink by infusion, i.e. by placing the bag in hot water, or by decoction, i.e. by boiling the bag in water.

Spray Drying: This is done through the use of modern equipment, the same method used in making spray-dried coffee. The herbs are boiled and concentrated, and finally placed in the spray-dry apparatus. The product is in a powder, easily dissolved in hot water.

Advantages and Disadvantages of Herb Cultivation

Herbs have different meanings to different people. To most laymen, herbs are synonymous with medicinal plants, while in this publication they are treated as culinary herbs, i.e. plants whose leafy and soft flowering parts are used as flavoring or seasoning materials. They are closely related to spices, which are non-leafy parts used for the same purpose as herbs. Both herbs and spices have been utilised by the Asian and other peoples throughout the world from time immemorial.

The recent impacts of urbanisation, industrialisation, and a 'back-to-nature' movement have led to increase use of herbs during the past decades. Never before have laymen expressed so much interest in using herbs, both as flavoring/seasoning materials, and for other purposes. This has stimulated cultivation of herbs for fresh consumption and for industrial processing. Before attempting to cultivate herbs, farmers should realise that herbs are special crops used on special occasions. Thus, there are both advantages and disadvantages in herb cultivation.

Advantages

The following advantages of herb cultivation are recognised:

Increasing demand: Industrialisation and the resulting urbanisation make it impossible for urban people to grow their own herbs while demand is increasing due to rises in living standard and increases in population.

Suitability of cultivation: Herbs can be cultivated in small areas for fresh-herb markets in the cities, or in large areas for industrial processing.

Ease in cultivation: Herbs are easy to grow from seeds, cutting or layering. Few pests and diseases are observed attacking herbs hence they rarely require pesticides.

Profitability: Herbs are high-value crops as compared to other commercial crops. They take less time (especially if they are annuals) and space than most other crops. They are easily marketed, either in the fresh or the dry stage. They can be sold retail or wholesale to local restaurants, at roadside stands, in market places or supermarkets, and to food processors or flavoring houses. Herbs can be processed and developed into more finished sale products such as herbal teas/drinks, herb vinegar, herb honey, herb jellies, etc. Proper packaging and labeling can enhance market opportunities.

Disadvantages

Herbs have also some disadvantages which farmers should know before cultivating them. They are:

Lack of information: Very little is known about herbs. Although numerous books are available, they either concentrate on medicinal rather than culinary herbs, or describe herbs of the temperate zones, many of which are not known in Asia.

Lack of improved cultivars: Except India where a number of improved herb varieties is available, most countries depend on primitive cultivars or introduced varieties that may not be productive.

Lack of agro-technology: Cultivation of herbs needs skill and experience that are not readily available to new growers.

Lack of reliable supply of seeds: Unlike other commercial crops where reliable supply of seeds is readily available, herbs, in most Asian countries, suffer from lack of such a service. Even if available, the service is of low quality standard.

Lack of registered pesticides: Since most herbs are produced for fresh consumption, the use of pesticides must be restricted to those that do not leave residues at the time of harvest. At present, pesticides specifically developed for herbs are not available.

Lack of reliable market: Markets for either fresh or processed herbs are either not available or not reliable.

2

Cultivation of Medicinal and Aromatic Plants

Medicinal plants play a central role, not only as traditional medicines used in many cultures, but also as trade commodities which meet the demand of often distant markets. The term "medicinal and aromatic plant" (MAP) is defined to cover the whole range of plants used not only medicinally sensu strictu but also in the neighbouring and often overlapping fields of condiments, food and cosmetics.

Demand for a wide variety of wild species is increasing with growth in human needs, numbers and commercial trade. With the increased realisation that some wild species are being over-exploited, a number of agencies are recommending that wild species be brought into cultivation systems. Cultivation can also have conservation impacts, however, and these need to be better understood. Medicinal plant production through cultivation, for example, can reduce the extent to which wild populations are harvested, but it also may lead to environmental degradation and loss of genetic diversity as well as loss of incentives to conserve wild populations.

The relationship between in-situ and ex-situ conservation of species is an interesting topic with implications for local communities, public and private land owners and managers, entire industries and, of course, wild species. Identifying the conservation benefits and costs of the different production systems for MAP should help guide policies as to whether species conservation should take place in nature or the nursery, or both.

Convention on Biological Diversity (CBD) and Ecosystem Approach

Since its adoption in 1992, the Convention on Biological Diversity (CBD) has strived to implement its three major goals: the conservation of biological diversity, the sustainable use of its components, and the fair and equitable sharing of the benefits from the use of genetic resources. Although MAP have not been explicitly on the agenda of the various CBD meetings, all three goals of the Convention are fully applicable to MAP resources.

The ecosystem approach is based on the application of appropriate scientific methodologies focused on levels of biological organisation which encompass the essential processes, functions and interactions among organisms and their environment. In April 2002, the CBD adopted the Global Strategy for Plant Conservation which provides a policy environment that is particularly well suited to addressing the conservation challenges for MAP in a coherent way.

Sustainability

As a base line element of the ecosystem approach it has to be recognised that humans, with their cultural diversity, are an integral component of ecosystems. In conceptual terms, the essence of sustainable development is expressed by the relationship between people and the ecosystem around it. This implies that ultimately one is entirely dependent upon the other. Human and ecosystem wellbeing need to be assessed together. A society is thought to be sustainable when both the human condition and the condition of the ecosystem are satisfactory or improving. The system improves only when both the condition of the ecosystem and the human condition improve.

Some Figures

MAP are used World-wide

The number of plant species which have at one time or another been used in some culture for medicinal purposes can only be estimated. An enumeration of the WHO from the late 1970s listed 21 000 medicinal species. However, in China alone 4 941 of 26 092 native species are used as drugs in Chinese traditional medicine, an astonishing 18.9 percent. If this proportion is calculated for other well-known medicinal floras and then applied to the global total of 422 000 flowering plant species, it can be estimated that the number of plant species used for medicinal purposes is more than 50 000.

Table 1: How many plants are used medicinally world-wide?

Country	Plant species	Medicinal plant species	%
China	26 092	4 941	18.9
India	15 000	3 000	20.0
Indonesia	22 500	1 000	4.4
Malaysia	15 500	1 200	7.7
Nepal	6 973	700	10.0
Pakistan	4 950	300	6.1
Philippines	8 931	850	9.5
Sri Lanka	3 314	550	16.6
Thailand	11 625	1 800	15.5
USA	21 641	2 564	11.8
Viet Nam	10 500	1 800	17.1
Average	13 366	1 700	12.5
World	422 000		52 885

We recognise, however, that certain plant families have higher proportions of medicinal plants than others. Good examples are the Apocynaceae, Araliaceae, Apiaceae, Asclepiadaceae, Canellaceae, Guttiferae and Menispermaceae. In addition, these families are not distributed uniformly across the world. As a consequence, not only do some floras have higher proportions of medicinal plants than others, but also have certain plant families a higher proportion of threatened species than others.

MAP Species are Traded

It is difficult to assess how many MAP are commercially traded, either on a national or even an international level. The bulk of the plant material is exported from developing countries while major markets are in the developed countries. An analysis of UNCTAD trade figures for 1981-1998 reflects this almost universal feature of MAP trade. Adding the volumes for the five European countries in this list (94 300 tonnes) marks the dominance of Europe as an import region. Germany ranks fourth and third as importer and exporter, expressing the country's major role as a turntable for medicinal plant raw materials worldwide.

Iqbal estimates that about "4 000 to 6 000 botanicals are of commercial importance", another source refers to 5-6 000 "botanicals entering the world market" (SCBD 2001). A thorough investigation of the German medicinal plant trade identified a total of 1 543 MAP being traded or offered on the German market. An extension of this survey to Europe as a whole arrived at 2 000 species in trade for medicinal purposes. Recognising the role of Europe as a sink for MAP traded from all regions of the world, it is a qualified guess that the total number of MAP in international trade will be around 2 500 species world-wide.

MAP are Threatened World-wide

To satisfy the regional and international markets, the plant sources for expanding local, regional and international markets are harvested in increasing volumes and largely from wild populations. Supplies of wild plants in general are increasingly limited by deforestation from logging and conversion to plantations, pasture and agriculture.

In many cases, the impact through direct off-take goes hand-in-hand with decline owing to changes in land use. Species favoured by extensive agricultural management like Arnica montana in central Europe go into decline with changes in farming practices towards higher nutrient input on the meadows. This requires habitat management as the key factor in managing species populations.

One of the goals of the IUCN Medicinal Plant Specialist Group is to identify the species that have become threatened by non-sustainable harvest and other factors. The enormity of this task is illustrated by the following estimate: According to Walter and Gillett, 34 000 species or 8 percent of the world's flora are threatened with extinction.

MAP are Under Cultivation

Many medicinal plants, especially the aromatic herbs, are grown in home gardens, some are cultivated as field crops, either in sole cropping or in intercropping systems and rarely as plantation crops.

In a survey carried out for the Rainforest Alliance, companies involved in trade and production of herbal remedies and other botanical products were asked what percentage of their material is from cultivated sources and what percentage from the wild. On average, companies reported that 60-90 percent of material was cultivated, with the remaining wild harvested. However, when asked about species numbers rather than volume of material, the figures are generally inverted. Lange and Schippmann state that of the 1 543 species traded in Germany, only 50-100 species (3-6 percent) are exclusively sourced from cultivation.

Of more than 400 plants species used for production of medicine by the Indian herbal industry, fewer than 20 species are currently under cultivation in different parts of the country. In China, about 5 000 medicinal plants have been identified and about 1 000 are more commonly used, but only 100-250 species are cultivated. In Hungary, a country with a long tradition of MAP cultivation, only 40 species are cultivated for commercial production. In Europe as a whole, only 130-140 MAP species are cultivated.

The number of MAP species currently in formal cultivation for commercial production does not exceed a few hundred world-wide. A global survey on the extent of MAP cultivation in terms of species, volumes and values would be highly desirable.

Wild and Cultivated Plants

Given the demand for a continuous and uniform supply of medicinal plants and the accelerating depletion of forest resources, increasing the number of medicinal plants species in cultivation would appear to be an important strategy for meeting a growing demand.

One explanation may be found in the observation that cultivated plants are sometimes considered qualitatively inferior when compared with wild gathered specimens. For instance, wild ginseng roots are 5-10 times more valuable than roots produced by artificial propagation. The reason is primarily cultural, as the Chinese community, which is the largest consumer group of wild ginseng, believes that the similarity in appearance of gnarled wild roots to the human body symbolises the vitality and potency of the root.

Cultivated roots lack the characteristic shape of wild roots and are therefore not as highly coveted by consumers. In Botswana, traditional medicinal practioners said that cultivated material was unacceptable, as cultivated plants did not have the power of material collected from the wild.

Scientific studies partly support this. Medicinal properties in plants are mainly due to the presence of secondary metabolites which the plants need in their natural environments under particular conditions of stress and competition and which perhaps would not be expressed under mono-culture conditions.

Active ingredient levels can be much lower in fast growing cultivated stocks, whereas wild populations can be older due to slow growth rates and can have higher levels of active ingredients. While it can be presumed that cultivated plants are likely to be somewhat different in their properties from those gathered from their natural habitats it is also clear that certain values in plants can be deliberately enhanced under controlled conditions of cultivation.

In general, in all countries, the trend is towards a greater proportion of cultivated material. The majority of companies, the mass-market, over-the-counter pharmaceutical companies as well as the larger herb companies, prefer cultivated material, particularly since cultivated material can be certified biodynamic or organic.

From the perspective of the market, domestication and cultivation provide a number of advantages over wild harvest for production of plant-based medicines:

(i) While wild collection often offers material adulterated with unwanted, sometimes harmful other plant species, cultivation provides reliable botanical identification.

(ii) Wild harvest volumes are dependent on many factors that cannot be controlled and the irregularity of supply is a common feature. Cultivation guarantees a steady source of raw material.

(iii) Wholesalers and pharmaceutical companies can agree on volumes and prices over time with the grower.

(iv) The selection and development of genotypes with commercially desirable traits from the wild or managed populations may offer opportunities for the economic development of the medicinal plant species as a crop.

(v) Cultivation allows controlled post-harvest handling and therefore

(vi) quality controls can be assured and

(vii) product standards can be adjusted to regulations and consumer preferences.

(viii) Cultivated material can be easily certified organic or biodynamic although certifiers are also presently developing wildcrafting standards.

However, domestication of the resource through farming is not always technically possible. Many species are difficult to cultivate because of certain biological features or ecological requirements (slow growth rate, special soil requirements, low germination rates, susceptibility to pests, etc.).

Economical feasibility is the main rationale for a decision to bring a species in cultivation but it also is a substantial limitation as long as sufficient volumes of material can still be obtained at a lower price from wild harvest. Cultivated material will be competing with material harvested from the wild that is supplied onto the market by commercial gatherers who have incurred no input costs for cultivation.

Low prices, whether for local use or the international pharmaceutical trade, ensure that few species can be marketed at a high enough price to make cultivation profitable. Domestication of a previously wild collected species does not only require substantial investment of capital but also requires several years of investigations.

On a time scale of sometimes many decades, the transition from wild harvesting to possible cultivation goes through various phases:

— *Discovery Phase:* At this point the demand can be met by wild harvest. Extractivism is done for local use or for barter with others.

— *Expansion Phase:* It is clear that the product is potentially useful and that demand is likely to increase. Harvest is done for local or regional sale and eventually for international markets. In general, species with naturally low densities are unlikely to become important sources of commercially large quantities.

— *Stabilisation Phase:* The species is unlikely to be attractive to growers unless prices are high enough and wild-harvested resources are scarce enough. However, desirable species may be grown on farm land and planted around settlements.

— *Decline Phase:* Prices increase with scarcity due to transport costs, search time and the long-distance trade. Wild populations will have to decline further before cultivation is a viable option. The trade is characterised by fluctuations in supplies, often to the extent of disrupting the trade balance. For slow growing species, if controls on collection are not strictly enforced, wild populations will be more seriously eroded before cultivated material is available.

— *Cultivation Phase:* Now, formal cultivation systems are developed and instituted. The plants are domesticated and incorporated in agroforestry systems sometimes for the benefit of small-scale farmers. If international market opportunities exist, commercial plantations are created with substantial investment and genetic selection, cloning, breeding and biotechnology may be applied. More resilient species may recover in their wild populations.

People Need Wild or Cultivated

Health Care Needs

There is a worldwide trend of increasing demand for many popular, effective species in Europe, North America and Asia, growing between 8–15 percent per year. Rapid urbanisation and the importance of herbal medicines in African health care systems stimulated a growing national and regional trade in Africa. Demand for medicinal plants also reflects distinct cultural preferences. In the USA, for example, only 3 percent of people surveyed had used herbal medicine in the past year, whereas in Germany, with a strong tradition of medicinal plant use, 31 percent of the over-the-counter products in pharmacies in 2001 were phytopharmaceutical preparations.

The level of herbal medicine use in most developing countries is much higher than this. While most traditional medicinal plants are gathered from the wild, these are not static health care systems, and introduced species are commonly adopted into the repertoire of plants used by African or South American herbalists. In many cases, herbal medicines can also be cheaper than western medicines, particularly where access to traditional healers is easier. Demand for traditional medicine continues in the urban environment even if western biomedicine is available.

Income Generation

Wild harvesting of medicinal plants is a chance for the poorest to make at least some cash income. Especially those people who do not have access to farm land at all depend on gathering MAP to earn at least some money. However, local people generally get a low price for unprocessed plant material.

Although income from Prunus africana bark sales is an important source of revenue to villagers in Madagascar, in some cases generating >30 percent of village revenue, the price paid to collectors is negligible compared to Madagascan middlemen. In Mexico, Hersch-Martinez found that medicinal plant collectors only received an average 6.17 percent of the medicinal plant consumer price.

Whether fruits, roots, bark or whole plants are involved, the potential yield from wild stocks of many species is frequently over-estimated, particularly if the effects of stochastic events is taken into account. As a result, commercial harvesting ventures based on wild populations can be characterised by a "boom and bust" situation where initial harvests are followed by declining resource availability.

Small-scale Cultivation

Small-scale cultivation, which requires low economic inputs, can be a response to declining local stocks, generating income and supplying regional markets. This can be a more secure income than from wild harvest which is notoriously inconsistent. For farmers that integrate MAP into agroforestry or small-scale farming systems, these species can provide a diversified and additional source of income to the family. Home gardens are increasingly a focus of medicinal plant propagation and introduction programmes intended to encourage the use of traditional remedies for common ailments by making the plant sources more accessible.

Large-scale Cultivation

As outlined by Leakey and Izac, large-scale cultivation has a number of socioeconomic impacts on rural people: "Commercialisation is both necessary and potentially harmful to farmers. It is necessary in that without it the market for products is small and the opportunity does not exist for rural people to generate income. A degree of product domestication is therefore desirable.

On the other hand, commercialisation is potentially harmful to rural people if it expands to the point that outsiders with capital to invest come in and develop large-scale monocultural plantations for export markets. Rural people may benefit from plantations as a result of available employment and hence off-farm income [...]. However, plantations may also distort market forces to their advantage, for example, by imposing low wages which will restrict the social and economic development of local people. The major beneficiaries of large-scale exports will probably be the country's elite and, perhaps, the national economy".

Also, those socially disadvantaged groups who actually depend on gathering MAP for their survival and cash income, may not have access to farm land at all and are therefore not able to compete with large-scale production of MAP by well-established farmers. Other limitations to the domestication approach include boombust and fickle markets that let farmers down when consumers turn their attention elsewhere.

Species and Ecosystems Require Wild or Cultivated Plants

Cultivation of medicinal plants is widely viewed not only as a means for meeting current and future demands for large volume production of plant-based drugs and herbal remedies, but also as a means for relieving harvest pressure on wild populations.

Booming markets with rapidly rising demands often have devastating effects on wild collected species. A closer look reveals that not all species are affected in the same way by harvesting pressures. The seven forms of rarity described by Rabinovitz make clear that a species which

(i) has a narrow geographic distribution,

(ii) is habitat specific, and

(iii) has small population sizes everywhere, is more easily over-harvested than species of any other pattern.

Secondly, the susceptibility or resilience to collection pressure varies among species owing to biological characters such as different growth rates (slow growing vs. fast growing), reproductive systems (vegetative or generative propagation; germination rates; dormance; apomixis) and life forms (annual; perennial; tree).

Species can be distinguished quite well in their susceptibility to over-collection if their life form and the plant parts collected are viewed together. Harvesting fruits from a long-lived tree presents a far lower threat to the long-term survival of the species than does collecting seeds from an annual plant. In the latter case, if the seed is gone the plant is gone.

In some cases the harvest impacts are more complex, e.g. with slow growing trees which reproduce from seed but only produce few, large fruits (example: Araucaria araucana, monkey puzzle tree). This will increase their susceptibility to over-harvest from low to medium or even high. A thorough summary of predictors of resilience or vulnerability to harvesting wild populations is presented by Cunningham.

Species most susceptible to over-harvest are habitat specific, slow growing and destructively harvested for their bark, roots or the whole plant. These species suffer most from harvesting and many of them have been seriously depleted, for example Prunus africana in West Africa, Warburgia salutaris in southern Africa and Saussurea costus in the Himalayas.

For threatened medicinal plant species cultivation is a conservation option because the constant drain of material from their populations is much higher than the annual sustained yield. If the demand for these species can be met from cultivated sources the pressure on the wild populations will be relieved. In these cases, the need for strict conservation of remaining populations, improved security of germplasm ex-situ and investment in selection and improvement programmes is extremely urgent as the example of Jaborandi (Pilocarpus jaborandi) in Brazil shows.

However, among the species that can be marketed at a high enough price to make cultivation profitable, only few are in the highest threat categories. Examples for threatened but cultivated species are Garcinia afzelii, Panax quinquefolius, Saussurea costus and Warburgia salutaris. With respect to economic viability many highly endangered MAP do not qualify for cultivation. This group of plants will enter cultivation only with the help of public domestication programmes.

For all other harvested MAP species the priority conservation option is sustainable harvest from wild populations, for a variety of reasons.

Let's imagine that a valuable medicinal plant is exploited by local collectors. A pharmaceutical company has domesticated and begun to cultivate the plant on a commercial scale. When the company no longer needs the wild-harvested material, local harvesters have to abandon the harvest and any incentive the local collectors might have had to protect the wild populations is gone. The domestication of MAP species has an environmental implication in the sense that it reduces the economic incentives for forest dependent people to conserve the ecosystems in which the MAP species occur.

If collectors and collecting communities can be involved in the development of propagation and management methods, the likelihood of their having an interest in protecting the wild populations from over-exploitation, particularly if these are understood to be the genetic resource "bank" for the domestic enterprises, will be greater.

Another aspect to consider is the genetic diversity of the species which is in demand. Long before non-sustainable harvest practices lead to extermination of a whole species, selection of favoured growth forms and concentration on certain harvesting areas which may hold certain ecotypes will lead to a degradation of genetic diversity of the wild populations. The same is true under domestication: Industry requirements for standardisation encourage a narrow genetic range of material in cultivation. Domestication will not achieve conservation of genetic diversity because a narrow group of high yielding individuals will be selected for planting.

Challenges for Sustainable Wild-collection

Sustainable harvest is increasingly seen to be the most important conservation strategy for most wild-harvested species and their habitats, given their current and potential contributions to local economies and their greater value to harvesters over the long term. The basic idea is that non-destructive harvests and local benefits will maintain population, species and ecosystem diversity.

Besides poverty and the break-down of traditional controls, the major challenges for sustainable wild-collection include: lack of knowledge about sustainable harvest rates and practices, undefined land use rights and lack of legislative and policy guidance.

Deficiency of Information on the Wild Resource

"The most important ingredient required to achieve a truly sustainable form of resource use is information". In reality, resource managers are always confronted with the lack of adequate information about the plants used, their distribution, the genetic diversity of wild populations and relatives and, above all, the annual sustained yield that can be harvested without damaging the populations. Research on the conservation and sustainable use of medicinal plants and their habitats has fallen far behind the demand for this globally important resource. Each species has unique ecological, socio-economic, health and cultural associations that must be understood. Model research approaches are feasible, model solutions are not.

Puzzles of Open Access Schemes

In many cases, access to the resource is open to everybody, rather than a limited access or private ownership. To make a living, commercial medicinal plant gatherers therefore "mine" rather than manage these resources. Open access schemes to harvestable plant population prevent rational and cautious use and make it difficult to adhere to quotas and closed seasons.

Lack of Legislative and Policy Support

Information on trade in MAP is scarce and data are rarely collected or published at a national level. Much production and consumption is at subsistence level and as a consequence the economic importance of these activities is largely under-estimated in government decision making regarding rural development, natural resource management planning and in government budget allocations. Therefore, national legislation and policies mostly fail to provide frameworks for a rational and sustainable use of wild resource.

Opportunities for governments to develop legislation to control and monitor harvest and trade of medicinal plant species and to consider conservation and sustainable use of medicinal plants as a priority in establishing protected areas have been greatly enhanced by two recent developments in international legislation: the addition of medicinal plant species to the Convention on International Trade of Endangered Species (CITES) and the entry into force of the CBD.

Future Trends and Solutions of Market Development

How will the market demand develop in the future? People in developing countries are already and will increasingly depend on medicinal plants as sources for their primary health care. An estimate by the World Health Organisation that more than 80 percent of the world's population relies solely or largely on traditional remedies for health care is frequently cited.

Also in the northern countries, use of medicinal plants is expected to rise globally, both in allopathic and herbal medicine. This upward trend is predicted not only because of population explosion, but also due to increasing popularity for natural-based, environmentally friendly products.

In general, the demand for medicinal plants and herbal remedies and especially its renaissance in the developed countries is driven by the following factors:

— increasing costs of institutional, pharmaceutical-based health care;

— interest of individuals, communities and national governments in greater self-reliance in health care;

— interest of communities and national governments in small and large-scale industrial development based on local/national biodiversity resources;

— increasing success in validating the safety and efficacy of herbal remedies;

— legislation improving the status of herbal medicine industry;

— renewed interest of companies in isolating useful compounds from plants;

— search for new drugs and treatments of serious and drug-resistant diseases;

— marketing strategies by the companies dealing in herbal medicine.

The limitations of cultivation as an alternative to wild harvest have been examined by Sheldon et al. in several case studies. Not with standing the level of interest in cultivation as a means for enhanced production and in a few cases as an effort to contribute to conservation of the resource, most medicinal and aromatic plant species will continue to be harvested wild to some extent. There is therefore a need to recognise and strengthen the role of local people in forest inventory, monitoring and impact assessment processes and to integrate non-timber product uses into forest management

Limiting the harvest to a sustainable level requires an effective management system and sound scientific information. The management system must include annual harvest quotas, consider seasonal or geographical restrictions and restriction of harvest to particular plant parts or size classes. In addition, clarification of the access and user rights to the resources providing MAP is part of the essential baseline information. Continuous monitoring and evaluation of the success is necessary to adapt the management strategy.

In many cases harvesting techniques need to be improved as the extraction of the roots or bark is often negatively affecting the recovery of the species or may even kill it. Collecting methods are often crude and wasteful, resulting in loss of quality and reduction in price.

Field-based methods have already been developed for sustainable harvest assessment and monitoring of non-wood forest products, resulting in the publication of research guidelines and predictive models.

Given that sustainable harvesting from the wild is difficult to achieve, certification standards can play a role to assure that a product meets certain standards of sustainability. Certification programmes related to natural resource use have mainly been developed for timber and agricultural products but they are presently being adapted for wild harvest of non-timber plants. Various schemes focus on different areas along the supply chain: production, processing, trade, manufacturing, marketing. Four categories of certification schemes have been identified to be of relevance for MAP products:

(i) forest management certification (e.g. Forest Stewardship Council FSC),

(ii) social certification (e.g. Fair Trade Federation FTF),

(iii) organic certification (e.g. International Federation of Organic Agriculture IFOAM), and

(iv) product quality certification.

The latter include parameters such as product identity, purity, safety and efficacy. The Good Harvesting Practices (GHP) developed for medicinal plants cover to some degree ecological aspects but need to be more clearly focussed on this aspect before they can make a meaningful contribution to ensuring sustainability.

3

Trends in Cut Flower Production

Floriculture has emerged as a profitable agri-business option the world over in recent years, particularly in the developing nations. World floriculture trade totals more than US$ 6,800 million (value of imports during 1995). The Netherlands alone account for nearly half of this trade and countries like Colombia, Italy, Denmark, USA, Belgium, Israel, Costa Rica, Germany, Canada, France, Spain, Kenya and Ecuador all have a share of more than US$ 100 million.

Asian countries are also entering the international trade and making their presence felt. Thailand, Malaysia, China, Singapore and India are leading the Asian contribution. However, while their present share of the trade appears insignificant, the growth of these countries in recent years has been very impressive. For cut flowers with a trade of US$ 3,700 million Thailand has a share of 2%. India which now has a share of only 0.3%, had a growth rate of 563% in the 1991-95 period. The other country with impressive growth rate has been Malaysia (259%) over the same period.

Among the growing markets (cut flowers, cut foliage, live plants and cuttings) Germany with a share of US$ 2,000 million was the world's biggest market, followed by

USA with a share of US$ 831 million. Japan, though seventh in ranking with a trade worth US$ 314 million, emerged as the fastest growing export market during the 1991-95 period indexing a growth of 162%. Consumption in these three countries are characterised by the domination of middle class population.

The world production of floriculture products at a very conservative estimate is valued at $ 20 billion. Japan with a production value worth US$ 3,736 million leads, and with the Netherlands (US$ 3,558 million) and the USA (US$ 3,270 million) account for nearly half of the world total.

In terms of total area in production, Asia and Pacific with an area of 134,000 ha cover nearly 60% of the total world area. A significant feature of the production area statistics is that in most Asian countries, only a small share of the area contributes to production for world trade (exports). This is because either only a small percentage of the area is under protected cultivation (as in China and India with large total areas) or the production is mainly for the domestic market (e.g. Japan).

Global production of floriculture products will continue to expand, especially in new producing countries in Asia, Africa and Latin America, though the traditional large producing countries like the Netherlands will also work towards higher productivity per worker and per unit area. In general, however, the trend in production of cut flowers which is going to be very labour intensive, will tend to be located in areas with lower labour and other production costs. The production strategies will require reduced costs so that the profitability improves. For the purchaser it is not important where the flowers come from. They are mainly interested in price, quality, freshness and variety.

For the growing flower exporting countries of Asia, one of the major barrier in increasing profitability is import duty. Around the world, duties are used to protect the local flower industry from unfair competition. The only exception being Japan, among the leading floriculture markets, which charges no duty as yet. But the total imports of cut flowers into Japan are less than 7% of their trade. It remains to be seen, if Japan will continue to have no non-tariff barriers or restrictions on trade, when China, assisted by Dutch technology and quality propagating material from around the world, increases its exports to Japan, particularly considering that China itself has a duty of nearly 53% (including 13% VAT).

For Asian countries, another hampering factor is the soaring cost of air freight, which may take away the competitive advantage of low cost land, labour and year round growing climate. Various international carriers no longer prefer flowers because they have alternative and more lucrative 'exotic' cargo such as computer chips.

The cut flower trade worth $ 3.7 billion is likely to increase by around 10% annually. The share of the developing countries of the total trade has consistently been around 20%. During the last five years there has been a consistent growth in the number of producing countries particularly among the developing nations in Asia, Africa and Latin America.

In general, the world supply is increasing more than demand, resulting in reduced prices per stem. With a highly perishable product like flowers, even a small over-supply can have a dramatic effect on prices. This will require a greater attention to be paid to identifying crops ideally suited for a particular location, as well as product diversification. One must critically analyse the strengths, weaknesses, opportunities and threats before embarking on expansion of the production/trade programmes. Greater attention to high value crops, though traded in lower volumes, will increasingly become more important.

Orchids, while not a major crop in most regions, have a special place in floriculture. Another crop demanding better attention in world trade has been Anthurium. Its world trade is second only to orchids among tropical flowers. Its popularity is increasing, particularly in Asia. Most Asian countries have potential for growing high value crops and need to devote larger areas mainly to meet the huge growing demand in Asian markets, where their freight rates will also be lower. Countries in Asia also require identification and commercialisation of certain native tropical plants to cater to niche markets in Europe and the USA.

As importers are trying to import from the cheapest country, the Asian growers could easily lose their large European market, in view of the strong competition from growers from Africa and Latin America, who may have lower freight costs and no or little import duty to pay. In addition the growers in Holland, with the advantage of producing right in the heart of the most developed market in the world, with low or no transportation cost or duty, would always enjoy the best marketing opportunity.

The Japanese and the US markets with relatively low import volume will continue to satisfy their supplies from nearby countries. The other growing Asian markets like Singapore, Hongkong, South Korea, etc. are hardly in a position to accept large volumes. One also needs to consider the emerging new markets in Middle East countries and Russia. The best option however, would be for all the major flower growing nations in the region to concentrate on developing their own domestic market.

With increasing consumer wealth, population concentration in major urban areas and improved living standards, India, China, South Korea among others appear to be potentially important flower consuming countries for the future. In these countries, as also in others, most flower consumption occurs around major holidays, like New Year, Valentine's Day, Memorial Day, Mothers' Day, Fathers' Day, Christmas, and in wedding seasons. The governments in these countries are making sincere efforts to develop the philosophy of people with regard to flowers through strong promotional efforts.

Foreign investors are being encouraged to help develop their floriculture industry through joint venture projects. All these efforts will hopefully lead to making available quality produce for the consumers. The governments will need to invest more on developing the required infrastructure, particularly for marketing like auction centres etc. (already being developed in Korea and India), and increase access to new and improved technologies and quality inputs. The efforts to develop the competitiveness in the export markets, through providing subsidies for capital investments, air freight etc. should also continue alongside.

With the growth of the flower industry the world over and increased exports, there is need to identify varieties that are most tolerant to the stresses of long shipping and storage. Consumers are seeking products which offer diversity in flower and leaf colour, good texture and form, as well as better performance in interior environments. Efforts to develop techniques which will reduce the costs of the ultimate product for the consumer will need to be strengthened, with identification of low cost production technologies in protected conditions and refinement and reorientation of imported technology to suit the requirements.

Public concern for the environment will require Asian growers to re-examine and change their irrigation, fertilisation and insect control practices. They will also have to reduce the use of pesticides and other chemicals, through effective IPM programmes and increased use of biological control methods. This is important due to the fact that, apart from environmental concerns, developing new pesticides is becoming expensive and governments in several countries are enforcing strict registration requirements for new pesticides, besides removing several pesticides from the market.

Development of the cut flower industry in the Asian region will thus require concerted efforts, not only on the part of the concerned government and growers of a country, but also joint programmes between countries of the region. In view of the commonality of problems and issues, sharing information on initiatives and developing effective networking systems among nations will lead to quicker and less expensive redressals.

Potential of Commercial Floriculture in Asia

The Asian flower industry has the potential to become the leading flower industry in commercial floriculture worldwide. Looking to the future, the Asian flower industry could rival, if not surpass, the size and scope of the European flower industry which presently dominates global commercial floriculture.

World Cut Flower Market Place

Forty years ago, demand for cut flower by consumers around the world was satisfied by local cut flower production. In Europe, per capita consumption was significant, and consumer culture required a large supply of cut flowers for gifts, occasions, and everyday use. As a result, cut flower production in Europe was sizable. Gradually as transportation systems developed throughout this region, it became possible to distribute cut flowers grown in southern areas of Europe to northern areas of Europe.

Consequently, the European flower industry began to extend its boundaries for cut flower production and along with this expansion grew the influence of the European flower industry. This background history could be considered the beginning of commercial floriculture as we know it today. When the world energy crisis occurred in 1973, the marketing plan for distributing cut flowers grown in different European countries to Holland for sale through the Dutch flower auction and back to markets throughout Europe became a significant production opportunity for southern European cut flower growers.

Increasingly larger quantities of cut flowers were grown in southern Europe to meet the demand for cut flower sales through Holland. Flower growers in the southern regions had a price advantage over growers located in northern regions because cut flower production was more expensive for northern growers during the winter season due to increased energy costs required to obtain quality flowers in controlled temperature greenhouses.

Then, competition for southern European cut flower growers intensified when Israeli cut flower growers, who were located further south entered the market with product to be sold through the Dutch flower auction. Israeli growers had the production advantage of being further south where they could produce cut flowers in open fields or plastic tunnels year round, eliminating most of the overhead expenses for greenhouses and heating systems. But in order to develop a potentially lucrative export cut flower industry for themselves, the Israelis needed to address limiting factors to their success. The two main limiting factors were transportation costs to Europe and a water shortage if production were to expand.

Solutions to these limiting factors were found for Israeli growers. In the case of transportation costs which offset growers cost advantage in terms of energy compared to growers in southern Europe, the government provided transportation subsidies which have reduced the costs to the growers to ship their cut flower product to Europe, thereby maintaining a competitive cost advantage over European growers. As for the water shortage, research on irrigation systems that would conserve water usage was applied to production systems for cut flowers.

Through the 1970's, the activities of the European flower industry had begun to influence cut flower production and sales beyond the borders of Europe. Cut flower sales through the Dutch flower auctions had gained a share of the United States market. This was achieved by promotion activities in the USA supported by the Holland Flower Council which encouraged Americans to purchase more cut flowers for gifts, occasions and everyday use, similar to consumer habits in Europe. Most of the flowers sold to the USA through the Dutch flower auctions are shipped to the USA by air through New York.

Simultaneously, Miami, USA, was being developed as a key import distribution base for cut flowers being grown in Columbia, South America and shipped north. This caused considerable competition for local cut flower growers in the USA. Manufacturers and suppliers from the European flower industry were quick to find opportunity in this situation. Not only were South American cut flower growers purchasing varieties from Europe but flower growers from the USA were persuaded to invest in production systems and equipment from Europe in hopes of becoming more efficient producers like the Dutch growers who had once faced competition from southern European growers. As a result, the United States flower industry owes a significant share of its growth in terms of promotion and sales and improved production systems to the influence of the European flower industry.

It is worthwhile to mention that the Israeli flower industry has become a formidable competitor of the European flower industry. Israeli cut flower producers ship significant quantities of product into the USA market via both New York and Miami. This compensates Israeli producers for the reduction in cut flower sales to the European market which is increasingly being supplied by flower growers from regions in Africa, especially Kenya. Also, Israelis have been successful in selling their production equipment and varieties to flower growers in other countries.

Continuing to advance in the 1980's, the European flower industry began seeking further opportunity and expansion in Asia by 1985. Japan's bubble economy was starting to inflate and discretionary income spending by the Japanese was rising. European flower imports made headway into the lucrative market in Japan. Within a few years, as economies in Korea, Taiwan, and Hongkong strengthened, the European flower industry moved into these markets with their cut flower exports as well.

Since the early 1990's, the European flower industry, as a worldwide leader in commercial floriculture, has been impacting the rest of Asia with cut flower imports from Holland and sales of flower varieties, production equipment, and technology for new production operations in Asia. Israeli cut flower producers, manufacturers, and suppliers have followed but, one step behind.

The main difference between the European flower industry and the Israeli flower industry is that the European flower industry enters their new markets by launching aggressive marketing campaigns which call attention to the quality and image of Dutch flowers. These campaigns stimulate demand by new consumers for their cut flower products. So far, the Israelis have not particularly created an image for end consumers of Israeli flowers. This difference is one of the factors which contributes to why the European flower industry is the world wide leader in commercial floriculture.

Initially, commercial floriculture production in southeast Asia was developed because of increasing need for low cost flowers by the European cut flower market place. European flower traders identified commercial floriculture production in southeast Asian countries as a source of supply. Ironically, Dutch auctions often served to re-distribute this product to the Japanese market. By the mid to late 1980's, Dutch importers/exporters had begun selling floriculture product in Japan. With economies expanding, the "little tigers", i.e. Taiwan, Korea and Hongkong were the next Asian targets with market needs for floriculture products from Europe and potentially from other Asian countries which could produce floriculture products less expensively.

The development of the commercial cut flower industry in Asia has been unlike that of Israel, African countries, south and central American countries. In the latter regions, cut flowers have been a product produced mainly for export with no thought of a potential domestic market. On the other hand, in Asia, whereas cut flowers were initially produced for export, the market potential has rapidly changed to include opportunities for supplying the local market as well. This unique development is on account of the rapid strengthening of economies in the region, high population densities, and the consumer perception which has been promoted heavily by the European flower industry that the use of fresh flowers in one's every day life represents an improved, quality lifestyle.

Today and in the future, the potential for commercial floriculture expansion in Asia, including production for domestic and export sales of cut flowers, is greater than ever before. The elements for success needed to transform the Asian flower industry into the worldwide leader of commercial floriculture have been implied by reviewing the evolution of the world cut flower market place.

Opportunities for Cut Flower Development

There is considerable opportunity in both export and domestic markets for cut flowers from southern Asia. In case of export markets, it seems reasonable to set short, middle, and long term goals with respect to entering these markets. Simultaneously, efforts to develop domestic markets within and among countries in southeast Asia should be undertaken.

Export Markets from Southeast Asia

Short - Term: Countries in Asia - Japan, Korea, Taiwan, Singapore

Already, cut flowers have been exported from southeast Asia to other countries in Asia, especially Japan, Korea, Taiwan, and Singapore. The advantages in marketing product to these countries is that they are relatively near, shipping requirements in terms of postharvest storage have been minimal, though this situation may not persist, and often contacts to find buyers have been made only through relative or previous business relationships. In the case of Japan, it is well known that custom regulations and quality requirements are among the strictest in the world, though the price fetched for product is high. This incentive is attractive despite the difficulty to achieve success in marketing cut flowers to Japan.

In terms of distribution to these countries, it is important to examine what are the practical transportation routes to these countries from southeast Asia. It could be that developing a transportation hub in Bangkok or Singapore would be convenient for shipping cut flowers from southeast Asian countries, including cut flowers from Yunnan province in China, to countries in northeast Asia. Numerous commercial flights already are in place regarding this possible shipping pattern for cut flowers from the region. The transportation concept of a shipping hub would imply that initially all cut flower production intended for export to northeast Asian destinations would first be shipped to the hub, e.g. Bangkok, and then re-distributed to specific destinations in Japan, Korea, Taiwan, and Singapore. This routing for northeast Asia could be extended to Australia and New Zealand.

This centralised shipping hub would require expanding postharvest handling and storage facilities at the airport terminal and increasing freight forwarding activities for cut flowers in Bangkok. At the same time, shipping procedures and storage requirements for cut flowers to be transported from southeast Asian countries would be streamlined by the ability to focus on transportation requirements to Thailand rather than an array of destinations. In the short term, this could serve as a way to shorten the time to improve distribution of cut flower product from all southeast Asian countries to these lucrative northeast Asian markets.

Mid-Term: Countries in North America - Canada and USA; Central Europe

At first consideration, it may seem unlikely that cut flowers exported from southeast Asia could penetrate the markets in North America or central Europe as these markets are already well supplied with cut flowers from other regions such as south and central America and southern Europe, Israel, and Africa. However, in the case of North America, the entry points for cut flower imports are mainly New York and Miami, leaving open the opportunity for establishing an import entry point in Vancouver or San Francisco, especially for cut flowers shipped from a Chinese port, i.e. Kunming. Recently, with the return of Hongkong to China, there has been an exodus of Chinese from Hongkong to Canada establishing residence in North America. Already the San Francisco area in the USA is home to many ethnic Chinese.

Due to the high density of Chinese in these two western entry points to North America, it is highly conceivable that Chinese flower importing companies could proliferate if the opportunity were available. This was certainly the case with Miami which already housed a large population from south and central America when flower imports from Columbia, Ecuador, and Costa Rica began to enter the USA market on the east coast.

Until now, domestic producers of cut flowers in California on the west coast of the USA have been holding up against imports from south and central America but, would surely retreat if low cost, high quality imports from southeast Asia via China were available. The reason for setting this export goal as "mid-term" is that airline shipping routes from China to these cities are not as well established at this time compared to routes from Bangkok to northeast Asian destinations.

It might be proposed to establish Kunming in China, which is currently being developed as the center for cut flower production in China, as the transportation hub for distributing flowers from southeast Asia to North America.

Another mid-term export goal would be to market flowers from southeast Asia to central Europe. This could be achieved, transportation wise also through Bangkok, which already has airline clearances for many central European cities, including Amsterdam.

Another factor in identifying mid-term export market goals is that both north American and central European markets are already well supplied with an abundance of quality, low cost flowers from other regions. These export markets are also both further away in distance so that shipping costs are not particularly an advantage. Marketing factors which will affect the attainment of these mid-term goals will likely be related to price and customer service. The capability for the Asian flower industry to initially offer these export market customers low prices, a full production mix, convenient and reliable shipping service will increase the chances of gaining market share in a market place which already has supply.

Long-Term: Countries in Eastern Europe

Lastly, there will be opportunities for marketing cut flowers from southeast Asia to Eastern Europe, particularly northeastern Europe. This would be another likely destination for flowers being shipped out of Kunming. At the moment, economies in northeastern Europe need to improve before significant quantities of cut flowers would be in demand, but by that time (5-10 years) it is likely that transportation routes will have expanded greatly from China to its neighbouring territory, northeastern Europe. As in the past, no doubt winter production of cut flowers in a southern area will surpass cut flower production in the north.

Domestic Markets within Southeast Asia

Similar to establishing export markets for cut flowers from southeast Asia, focusing on improving transportation is a key factor to establishing distribution into domestic markets within and among southeast Asian countries. It is proposed that distribution between these countries may be improved if a transportation hub system were enacted. Also, in the case of developing domestic markets, it is likely that efforts to add or improve wholesale markets in major cities would facilitate greater distribution of cut flowers within these countries.

Cut Flower Development in Asia - Production

Opportunities for cut flower development in Asia with respect to production are not location dependent, meaning that any area of southeast Asia should be able to capture significant cut flower production opportunity if pursued. In present and future commercial floriculture markets, production opportunities exist if the goal to reliably produce high quality product in consistent quantities can be attained.

To reliably produce high quality cut flowers in consistent quantities requires optimum production management. Production management strategies for implementation and training of personnel will be different for different types of production operations. Varying types of production operations might be commercial cut flower companies, large or small; government directed agricultural farming communities; or rural development programmes managed by public or private funds. For each of these types of production operations, dissemination of pertinent information needed to manage production may require different strategies. Therefore, a key factor to realising production opportunities is education. Ideally, educational resources should be developed and supported by public and private sectors.

An example of educational resources which could be critical to the success of implementing optimum production management for cut flower enterprise in Asia would be a training center/demonstration showcase of production technology adapted to local growing conditions for specific cut flower crops. This educational institution or organisation could facilitate dissemination of information to management at cut flower operations or provide training of technical personnel involved in cut flower operations.

Enlisting the support of floriculture professionals from outside and within Asia to coordinate activities in research and education for the purpose of furthering floriculture development in Asia would have significant impact on the pace and scope of cut flower development in Asia. Formally organised, it would be possible to solicit funds through international agencies to support an institution set up for this purpose, i.e. International Center for Floriculture in Asia.

Departments related to production and marketing management for cut flower crops would be defined in order to reliably produce high quality cut flowers in consistent quantities regardless of the location in southeast Asia. Such departments could be:

— Cultural production, information and management.

— Postharvest production, information and management.

— Equipment use, design, and engineering.

— Genetic improvement of cut flower varieties.

— Consumer marketing.

If sufficient attention and support were given to attaining the goal of reliably producing high quality cut flower product in consistent quantities among Asian countries, then it would be reasonable to expect that the Asian flower industry would soon overtake any other region's flower industry, including the European flower industry, in production. Moreover, given the rapidly increasing rate of spending among Asian consumers for cut flowers, it is also reasonable to expect that the Asian flower industry will soon surpass consumption rates for cut flowers compared to that of other regions.

Leadership in Commercial Floriculture

It is exciting to discuss how and why the Asian flower industry has the potential to not only "take the lead" in volume and consumption of floriculture products but, to "become the leader" of commercial floriculture worldwide. This may suggest the most important marketing and production opportunities for cut flower development in Asia.

Once again, in review of activities of the European flower industry, it can be said that the European flower industry has led the way in floriculture production, research, floriculture education and training programmes, development of floriculture postharvest handling and storage technology; floral marketing strategies; grades and standards for floral product; and transportation systems for distribution of flowers. Countries within the European region and outside of the region depend on product innovation from Europe to yield new flower varieties. Innovative floriculture production systems have been promoted and exported to other regions of the world. Even floral design concepts have been defined by the European flower industry.

The European flower industry exports more than just flowers. The European flower industry exports flowers, equipment and technology to produce flowers, marketing programmes to increase consumer purchases, and with those marketing programmes, cultural traditions and attitudes about flowers which includes floral design, uses of floral products, i.e. decoration, gifts, holidays, and appreciation for flowers.

Celebration of western holidays and the purchases of flowers for these holidays have become a main theme of floral marketing for consumers in Asia. Two examples of this are cut roses for Valentine's Day and cut carnations for Mother's Day. This type comprehensive involvement with floriculture from disseminating the use of production equipment and technology to influencing consumer purchases of flowers by promoting western holidays associated with floral purchases has established the European flower industry as the leader of commercial floriculture worldwide.

There are many opportunities for marketing cut flowers from Asia. In addition to developing a new transportation concept for improving global distribution of cut flowers produced in Asia, there are opportunities in developing marketing strategies which would boost global consumption of flowers produced in Asia. First of all, unifying the image of flowers produced in Asia to create an identity of Asian floriculture would be one step. Then, this could be implemented by introducing an Asian calendar of floral holidays which indicated particular flowers and how they are used in association with these holidays. Even if certain flowers are not presently used in association with a particular holiday or celebration in Asia, there is no reason that floral associations from each country could not invent the concept.

For example, in China, on a certain day in the autumn, the moon and all things associated to be round like the moon are appreciated. Millions of round cakes, labelled "moon cakes" are sold to consumers. Why could not this holiday become a commercial opportunity for the Asian flower industry and pom pom mums become the flower of choice for this day as a gift item along with "moon cakes". The same can be said for Americans and Europeans. In Vietnam, flower consumption is auspicious on the 1st and 15th of the month in compliance with Vietnamese traditions. This seems like another convenient way for the Asian flower industry to boost cut flower consumption all over the world twice a month if there were an Asian calendar of floral holidays reminding consumers that it was that time of the month to buy flowers.

Imbedded in this type of floral promotion strategy is the opportunity for the Asian flower industry to develop new varieties of flowers which would characterise these holidays. For example, in case of "Moon Day" not just any type of pom pom mum would qualify, but a particular colour, size, shape, etc. could be developed.

Given the cultural differences among Asian countries, the challenge would be to incorporate as many different cultural traditions representing the different Asian cultures as part of the coordinated Asian flower industry effort. This approach would allow each country to capitalise exclusively on a particular holiday, boosting production and sales for that country according to the holiday celebrated. Once the mission of developing an Asian flower industry was accepted, it would be certain that marketing ingenuity from within Asian flower associations would take over and demand for cut flower products from Asia would soar.

Opportunities in production of cut flowers in Asia would go beyond just the need to increase the rate of production. New production systems for efficiency or environmental protection could be developed which would be adapted to conditions of flower production in Asia. Research priorities at flower research institutions throughout Asia could be directed towards developing innovative solutions which supported global sustainability. Then, in a technology turnaround, it would be countries in the west which would be sending exchange staff to Asia to study new flower production systems.

Novel engineering or design applications in equipment and supplies which would be used by producers of flowers for the Asian flower industry could be patented and exported to flower producers in other countries outside of Asia. Endless opportunities. Endless opportunities. The establishment of an Asian flower industry could direct the unlimited opportunities for development of cut flower production in Asia and development of worldwide flower markets for cut flowers from Asia.

Cut Flower Production in India

India has a long tradition of floriculture. References to flowers and gardens are found in ancient Sanskrit classics like the Rig Veda (C 3000-2000 BC), Ramayana (C 1200-1300 BC), Mahabharata (prior to 4th Century BC), Shudraka (100 BC), Ashvagodha (C 100 AD), Kalidasa (C 400 AD) and Sarangdhara (C 1200 AD). The social and economic aspects of flower growing were, however, recognised much later. The offering and exchange of flowers on all social occasions, in places of worship and their use for adornment of hair by women and for home decoration have become an integral part of human living. With changing life styles and increased urban affluence, floriculture has assumed a definite commercial status in recent times and during the past 2-3 decades particularly.

Appreciation of the potential of commercial floriculture has resulted in the blossoming of this field into a viable agri-business option. Availability of natural resources like diverse agro-climatic conditions permit production of a wide range of temperate and tropical flowers, almost all through the year in some part of the country or other. Improved communication facilities have increased their availability in every part of the country. The commercial activity of production and marketing of floriculture products is also a source of gainful and quality employment to scores of people.

Present Situation

Inspite of the long and close association with floriculture, the records of commercial activity in the field are very few. The information on the area under floriculture and the production generated is highly inadequate. As commercial floriculture is an activity which has assumed importance only in recent times, there are not many large farms engaged in organised floriculture. In most part of the country flower growing is carried out on small holdings, mainly as a part of the regular agriculture systems.

Flower Production Areas

The estimated area under flower growing in the country is about 65,000 hectares. The major flower growing states are Karnataka, Tamil Nadu and Andhra Pradesh in the South, West Bengal in the East, Maharashtra in the West and Rajasthan, Delhi and Haryana in the North. It must, however, be mentioned that it is extremely difficult to compute the statistics of area in view of the very small sizes of holdings, which very often go unreported. This perhaps would be the reason for unrealistically small areas reported for floriculturally active states like Maharashtra, Uttar Pradesh and Madhya Pradesh.

More than two thirds of this large area is devoted for production of traditional flowers, which are marketed loose e.g. marigold, jasmine, chrysanthemum, aster, crossandra, tuberose etc. The area under cut flower crops (with stems) used for bouquets, arrangements etc. has grown in recent years, with growing affluence and people's interest in using flowers as gifts. The major flowers in this category are rose, gladiolus, tuberose, carnation, orchids and more recently liliums, gerbera, chrysanthemum, gypsophila etc.

Table 1. Area Under Flower Production in India

State	Area (ha.)
Karnataka	19,161
Tamil Nadu	14,194
West Bengal	12,285
Andhra Pradesh	5,933
Maharashtra	3,356
Rajasthan	1,985
Delhi	1,878
Haryana	1,540
Madhya Pradesh	1,270
Uttar Pradesh	1,000
Others	2,166
Total	64,768

The production of flowers is estimated to be nearly 300,000 metric tonnes of loose flowers and over 500 million cut flowers with stem. In the case of production also, the estimates could be at variance from the actual figures as some of the flowers like rose, chrysanthemum, and tuberose are used both as loose flowers and with stem.

It may be mentioned that almost all of the area reported here is under open field cultivation of flowers. Protected cultivation of flowers has been taken up only in recent years for production of cut flowers for exports. The estimated area in production is about 200 hectares, which is likely to increase to over 500 hectares by the year 2000.

Recognising the potential for low cost production for export, in view of cheap land, labour and other resources, several export oriented units are being set up in the country. These projects, located in clusters around Pune (Maharashtra) in the West, Bangalore (Karnataka) and Hyderabad (Andhra Pradesh) in the South, and Delhi in the North, are coming up in technical collaboration with expertise mainly from Holland and Israel. More than 90 percent of these units are for rose production, on an average size of 3-hectare farm, while some projects for orchid, anthurium, gladiolus and carnation are also being set up. Nearly one third of over 200 proposed projects, have already commenced production and export.

Major Cut Flower

Rose is the principal cut flower grown all over the country, even though in terms of total area, it may not be so. The larger percentage of the area in many states is used for growing scented rose, usually local varieties akin to the Gruss en Tepelitz, the old favourite to be sold as loose flowers. These are used for offerings at places of worship, for the extraction of essential oils and also used in garlands.

For cut flower use, the old rose varieties like Queen Elizabeth, Super Star, Montezuma, Papa Meilland, Christian Dior, Eiffel Tower, Kiss of Fire, Golden Giant, Garde Henkel, First Prize etc. are still popular. In recent times, with production for export gaining ground in the country, the latest varieties like First Red, Grand Gala, Konfitti, Ravel, Tineke, Sacha, Prophyta, Pareo, Noblesse. Virsilia, Vivaldi etc. are also being grown commercially.

Gladiolus is the next most important cut flower crop in the country. Earlier it was considered a crop for temperate regions and its growing was restricted to the hilly areas, particularly in the north eastern region, which still continues to supply the planting material to most parts of the country. However, with improved agronomic techniques and better management, the northern plains of Delhi, Haryana, Punjab, Uttar Pradesh, as well as Maharashtra and Karnataka have emerged as the major areas for production of gladiolus.

Tuberose, a very popular cut flower crop in India is grown mainly in the eastern part of the country i.e. West Bengal, and also in northern plains and parts of south. Both single and double flower varieties are equally popular. Tuberose flowers are also sold loose in some areas for preparing garlands and wreaths.

The other main cut flower item is orchid. Its production is restricted mainly in the north-eastern hill regions, besides parts of the southern states of Kerala and Karnataka. The main species grown are Dendrobiums, Vanda, Paphiopedilums, Oncidiums, Phalaenopsis and Cymbidiums.

Among the traditional crops grown for loose flowers, the largest area is under marigold, grown all over the country. In most parts of the country only local varieties are grown for generations. African marigolds occupy more area as compared to the small flowered French types. Jasmine flowers in view of its scent are also very popular as loose flowers and for use in garlands and Veni (ornament for decoration of hair by women).

The major areas under this crop are in Tamil Nadu, Karnataka in South and West Bengal in East. The varieties are mainly improved clones of *Jasminum grandiflorum, J. auriculatum and J. sambac*. The chrysanthemum, particularly the white varieties are much in demand as loose flowers during the autumn period of October-December when other flowers like jasmine, tuberose are not available for use in garlands etc. Among other traditional flowers grown in large areas are crossandra in southern states of Tamil Nadu, Karnataka and Andhra Pradesh and aster in Maharashtra.

Research Support on Floriculture

Research work on floriculture is being carried out at several research institutions under the Indian Council of Agricultural Research and Council of Scientific and Industrial Research, in the horticulture/floriculture departments of State Agricultural Universities and under the All India Coordinated Floriculture Improvement Project with a network of about twenty centres. The crops which have received larger attention include rose, gladiolus, chrysanthemum, orchid, jasmine, tuberose, aster, marigold etc.

The thrust till recently had been on crop improvement, standardisation of agro-techniques including improved propagation methods, plant protection and post harvest management. In view of the fact that most of the cut flower production is being done under open field conditions, the research efforts generally relate to open cultivation. In recent years, however, technologies for protected cultivation and tissue culture for mass propagation have also received attention. A large number of varieties suitable for cut flower use, as well as garden display have been developed. Production technology, particularly the agronomic requirements and control methods for important diseases and insect pests have also been developed. Contribution by the private sector in research activities in floriculture is negligible.

Requirement of Planting Material

The requirement of planting material to cater to the large

area under flower crops, is largely met from domestic production. Since efforts to set up large commercial farms generally suffered due to lack of quality planting material in sufficient quantities, this aspect has received greater attention in recent years in the breeding centres, which are producing sufficient quantity of planting material. Most of the nurseries propagating planting material are in the private sector.

In the absence of any mechanism to register nurseries, it is very difficult to ascertain their exact number, but at a very conservative estimate there are more than 100,000 nurseries, spread out all over the country, producing seeds and other planting materials for flower growers. The states with larger numbers of nurseries include Maharashtra, West Bengal, Karnataka and Tamil Nadu. Most of the nurseries are small, with little or no improved facilities like mist propagation unit, green houses/net houses etc. For meeting the demand of flower seeds, several large seed companies have production units in Punjab, Himachal Pradesh and Jammu & Kashmir in the North, Karnataka in the South and West Bengal in the East.

A few of the leading multinational seed companies have tied up with local seed companies or producers for custom production of seeds of their varieties. In the case of bulbous plants, most of the planting material is produced in the north eastern hilly regions of West Bengal (Kalimpong) and Sikkim, though for some crops, it is also produced in hilly regions of northern India. The introduction of a revised seed policy by the government of India in 1989 has enabled unrestricted introduction of many new and superior varieties into the country, increasing the variety in the floral basket.

Tissue culture has, in recent years, been recognised as an important tool in agriculture development. With its diverse climatic zones and qualified manpower, India is well placed to exploit the benefit of tissue culture based applications to floriculture crops. Most popular application of tissue culture has been micropropagation using in vitro technique for mass multiplication of planting material.

Tissue culture plants of ornamentals have found ready acceptance by the commercial growers and their production increased significantly from 130 million plants in 1985-86 to 680 million in 1994-95. At present 30 commercial tissue culture units with annual capacities of 0.5 to 15 million plants each are in operation, resulting in total capacity of about 110 million plants. While most of it is exported, a small percentage of cut flower crops like carnation and gerbera are finding good market within the country.

Marketing of Cut Flowers

Marketing of cut flowers in India is very unorganised at present. In most metropolitan cities, with large market potential, flowers are brought to wholesale markets, which mostly operate in open yards. A few large flower merchants generally buy most of the produce and distribute them to local retail outlets after significant mark up.

The retail florist shops also usually operate in the open on-road sides, with different flowers arranged in large buckets. In the metros, however, there are some good florist show rooms, where flowers are kept in controlled temperature conditions, with considerable attention to value added service. The government is now investing in setting up of auction platforms, as well as organised florist shops with better storage facilities to prolong shelf life.

The packaging and transportation of flowers from the production centres to the wholesale markets at present is very unscientific. The flowers, depending on the kind, are packed in old gunny bags, bamboo baskets, simple cartons or just wrapped in old newspapers and transported to markets by road, rail or by air. The mode of transportation depends on the distance to the markets and the volume. Mostly, flowers are harvested in the evening time and transported to nearby cities by overnight trains or buses.

In recent years, the government has provided some assistance for buying refrigerated carriage vans. A large number of export oriented units have built up excellent facilities of pre-cooling chambers, cold stores and reefer vans and their produce coming for domestic market sales are thus of very good quality and have longer vase life and command higher price. The government programmes for floriculture development include creating common facilities of cool chain in large production areas to be shared on cooperative basis. Formation of growers' cooperatives/associations are being encouraged.

In view of the unorganised set up, it is difficult to estimate the size of flower trade, both in terms of volume and value. A study conducted in 1989 estimated the trade to be worth Rs. 2050 million. It is in the period of the last five years or so that this business has really boomed in India, which is reflected in the number of new florist outlets in all cities and increase in the public's purchase of flowers as gifts. This would put the current trade at several times the earlier estimate. A recent study of Delhi market alone put the value of flowers traded on wholesale as Rs. 500 million.

The loose flowers (traditional crops like marigold, jasmine etc.) are usually traded by weight. The average price of different flowers in major markets varies considerably depending on the period of availability.

Table 2. Average Market Price for Major Flower Crops

Flowers	Unit	Price (US$1 = Rs.40) Rs./kg or doz or each stem
Marigold	kg.	3-60
Jasmine	kg.	15-150
Crossandra	kg.	20-120
Chrysanthemum	kg.	5-25
Tuberose	kg.	5-30
Rose	kg.	6-60
Gladiolus	doz.	20-75
Carnation	doz.	30-75
Gerbera	doz.	36-75
Orchids	each stem	10-45
Liliums	each stem	10-45
Anthuriums	each stem	15-45

The net returns to the growers depend on the packaging and transportation costs. The cut flowers with stem have a limited overall market in terms of volume. The share of cut flowers has almost doubled from 30 to 60% in the last decade. The value of cut flower export from India has increased twenty five fold during the last five years. With more export oriented units coming into operation, exports are likely to grow further in the coming years.

The major share of the export trade is for roses, in addition to orchids, gladiolus etc. The major markets are Europe (Holland, Germany and U.K.) and Japan. The exports of roses to Japan, have really picked up in the three years from Rs. 360 million in 1993-94 to Rs. 6090 million in 1995-96. As per the estimates for 1996-97, India has been the largest supplier of roses to Japan (volume wise).

Table 3. Export of Floriculture Products from India (Rupees in Million)

Item	1991-92	1992-93	1993-94	1994-95	1995-96
Cut Flower Fresh	3.99	10.90	9.96	29.98	100.33
Dried Flowers	64.35	74.31	109.21	195.78	364.56
Live Plants	40.15	30.56	30.52	60.43	81.48
Dried Plants	23.99	25.45	23.72	10.35	35.83
Bulbs, Tubers etc.	12.06	7.83	14.95	11.83	19.21
Total	144.54	149.05	188.36	308.37	601.41

Cut Flower Production Development

The availability of natural resources like favourable and diverse climatic conditions permit production and availability of a large variety of flower crops round the year. Cheap labour leads to reduction in production costs, increasing access of the consumer to good quality flowers at affordable prices, besides increasing competitiveness in the export markets. Being a new concept in the agri-business, it took some time for scientific commercial flower production to take roots, but with the appreciation of its potential as an economically viable diversification option, its growth is slowly stabilising. The government also has, during the last few years, recognised floriculture as an important segment for developmental initiatives.

Model Floriculture Centres being set up in 11 major production zones, to serve as focal units for development in the region, have a mandate of making available quality planting material, new/improved production technologies and also to provide training in production and post harvest management. There are also special government programmes for area expansion in floriculture with state assistance. The National Horticulture Board, a major developmental agency for horticulture, also makes available finances as soft loan for setting up integrated projects for production and marketing. As mentioned earlier, the government is investing in improving the infrastructure for marketing in the domestic sector.

Production of cut flowers for exports is also a thrust area for support. The Agricultural and Processed Food Products Export Development Authority (APEDA), the nodal organisation for promotion of agri-exports including flowers, has introduced several schemes for promoting floriculture exports from the country. These relate to development of infrastructure, packaging, market development, air freight subsidy etc. The 100% Export Oriented Units are also given benefits like duty free imports of capital goods. All these efforts indicate the government's commitment for improving the sector and creating a positive environment for entrepreneurship development in the field.

Constraints in Flower Production

Being a new concept, the requirements of scientific and commercial floriculture is not properly understood in the country. The developmental initiatives of the government have to keep in mind the low knowledge base, small land holdings, unorganised marketing and poor infrastructural support.

While long experience of flower growing in the open field conditions enable sufficient flower production for domestic markets, the quality of the produce, in view of its exposure to various kinds of biotic and abiotic stresses, is not suitable for the ever growing export market. The production technology for flowers under protected environment of green houses needs to be standardised. There is hardly any post harvest management of flowers for the domestic market.

Availability of surplus flowers from exports for sale in the domestic market, has increased the appreciation of quality produce and the demand for good quality flowers is increasing. With the introduction of new varieties of crops in the country, facilities for generating their planting material for large scale production need strengthening. Special attention needs to be paid to strengthen the marketing infrastructure like organised marketing yards, auction platforms, controlled condition storage chambers etc.

Greater research efforts are also needed for integrated pest management, development of location specific package of practices for traditional flowers, value addition to traditional flowers etc. The initial cost and availability of finance is a critical matter in the development of large commercial projects requiring heavy investments. More options for developmental finance, such as the soft loan scheme of the National Horticulture Board need to be identified. In the initial years of commercial floriculture development, the governmental support in terms of subsidies etc. needs special attention.

The potential for growth of export market is always linked to the strength of domestic market - its capacity to absorb surplus and over production, and quality consciousness of consumers. Though we have a large domestic market, the marketing system and facilities need to be modernised. The production for exports at present has suffered due to a few constraints. While growers have been successful in producing world class quality at low cost, high air freight rates, low cargo capacity available, imposition of import duties, inadequate export infrastructure etc. have reduced their competitiveness.

There is also a shortage of trained manpower to handle commercial floriculture activity. The demands of the growing export oriented industry would require adequate attention to be paid for human resource development, particularly at the supervisory level.

Table 4. Cut Flower Exports from India

Country	1993-94	1994-95	1995-96
Japan	322.50	8255.58	35932.56
Netherlands	1004.61	9102.49	24799.90
U.S.A.	1175.38	2495.21	17652.50
Germany	957.61	2538.63	9256.00
U.K.	1420.93	1113.78	3345.86
U.A.E.	2120.40	3388.19	2459.29
Italy	210.28	164.96	2200.49
Hongkong	730.02	903.15	1504.02
Singapore	78.45	437.37	1190.54
Nepal	11.86	36.09	292.64
Kuwait	24.22	5.00	274.13
Saudi Arabia	413.98	169.52	272.77
Switzerland	136.96	258.20	242.83
Hungary	-	286.69	181.82
Thailand	-	86.49	177.82
Australia	-	-	132.89
Russia	368.46	20.96	119.85
Others	988.68	719.39	9293.82
Total	9964.34	29981.60	109329.73

Flower Forcing for Production

Flower forcing is an operation or treatment to the plant, after it reaches the ripeness-to-flower stage, in order to stimulate it to flower at a specific date (e.g. on New Year's day), or during off-season period. The flowering date/period may be earlier or later than the normal date/period of flowering. The goals of flower forcing are off-season production and specific-date production.

Cut flowers which are available during the normal season are plentiful, thus fetching a low price. Sometimes the farmers have to sell their produce even at a loss. In some cases, flowers which could not be sold are either left on the plants or are spoiled after being harvested.

Thus, it would be beneficial for farmers to produce cut flowers during the off-season period to obtain higher price, although the inputs may be higher.

Similarly, the demand for cut flowers is generally very high during certain occasions such as New Year, Christmas, Mother's Day, Memorial Day, Valentine Day, Graduation Exercise Day, etc. Thus, it will be to the farmers' advantage if they can produce cut flowers to be available on these specific dates.

The objectives of forcing plants to flower during off-season or at certain specific dates are:

— To avoid surpluses of in-season cut flowers,

— To avoid wastage or spoilage of surplus cut flowers,

— To avoid danger of epidemics,

— To distribute employment throughout the year,

— To increase farmers' income,

— To reduce imports and trade deficit,

— To satisfy customers at the time of their needs.

Physiology

The process of flowering follows the following sequence:

Proper stimuli (temperature & photoperiod).

Ripeness-to-flower stage — Vernalin — Florigin — Flower.

Factors affecting flowering are photoperiod, temperature and humidity.

Flowering behaviour of plants is controlled by seasonal changes. There are two types of flowers with respect to the seasonal effect on flowering:

— Little influence of seasonal changes: e.g. roses, marigold, chrysanthemum, heliconia, etc.

— Great seasonal influence: e.g. jasmine, dendrobium orchids, etc.

Seasonal factors can be of various types viz.

— Photoperiodic influence: This includes short day plants, which are temperature and humidity influenced; also long day plants, which are also temperature and humidity influenced.

— Temperature influence: This includes low temperature requiring plants, which are photoperiodic and humidity influenced; as also, high temperature requiring plants which are also photoperiodic and humidity influenced.

— Humidity influence: Including low and high humidity requiring plants.

Flower Forcing Operation

Flower forcing can be achieved by adjusting the factors effecting flowering behaviour, viz. photoperiod, temperature and humidity. Chemical flower forcing can be done through the application of fertilizers, used both for retarding and stimulating flowering; and plant hormones, including gibberellin, growth retardants and growth inhibitors. Mechanical flower forcing is achieved by operations such as, pruning, leaf trimming, ringing, budding or grafting, smoking, low-temperature storage and breaking dormancy.

Some Cut Flowers in Thailand

Dendrobium Orchids

Flowers are available all the year round, but more profuse

in rainy season and less profuse in winter (which happens to be the time when the demand is high). Pinching-off flower buds in August/September (to save nutrients for later blooming) and application of special fertiliser (high in P & K) during October/November, 3 to 4 times, induces blooming from November onwards for December/January harvest.

Siam Tulip

Seasonal flowers are produced during rainy season (June to August), requiring longday condition. No flower develops after September when shortday condition commences. Above ground parts wither and die down, and rhizome enters dormancy period until next rainy season.

Providing additional light breaks dormancy. Most effective is 3 hours of light in the middle of the night. Should be started soon after daylength is shortened (21 September). In this way, plants will continue to produce flowers all the way up to the New Year day, provided enough humidity and nutrients are given.

Marigold

Flowers are available round the year. It is a day neutral plant which takes 60 - 70 days from seeding to harvest. Timing of flowering can be maintained by fixing the date of seeding 60 to 70 days prior of harvest date. The recommended period is 65 days. For example, if the flowers are to be harvested for New Year, seeding should be done on 27 October, transplant on 6 November and pinching on 22 - 24 November. Flower buds emerge by 5 December, and blooming starts from 25 December to 5 January.

Lotus

Flowers are available all the year round, but the plant needs standing water. Low temperature during winter reduces the amount of blooming. Thus, it is recommended that the level of water be reduced to 50 cm in order to raise water temperature. In this way, blooming will be the same as in summer, both in terms of amount and size of flower.

High temperature during the summer speeds up the growth of lotus plant and the blooming time, but flower size remains small. Thus, water level should be raised up to 75 cm (from original 50 cm level in the winter). This way, water temperature will be reduced and the amount of bloom and size of flower will be maintained. During the rainy season, due to the addition of rain water the level of water may increase, thus it is recommended that the level of water be maintained at 50 cm in order to accommodate additional rain water. The blooming will be maintained as normal.

Chrysanthemum

It is a short day plant, with critical value of 14.4 hours. Thus, it will bloom all the year round under Thai conditions, having maximum daylength of 13.3 hours in June. Daylength can be extended by giving artifical light after sunset for about 3 hours during the early stage of growth to keep seedlings in vegetative stage until one month prior of the planned harvest date.

For example, if the planned harvest date is New Year, cutting should be made in September and transplanted to the growing plot for rooting to occur. Seedling is kept under light regime of more than 14.5 hrs by providing artifical light (100 w incandescent bulb) until 1 December (seedlings should be at least 30 cm high). These will bloom on 1 January.

However, as chrysanthemum blooms profusely during the period of low temperature which commences in December, it fetches a low price in the market even during the time of Christmas or New Year. Thus, some farmers avoid producing flowers during such period but shift it to the summer. The problem is that the temperature during summer is quite high and not optimum for chrysanthemum growth.

The most suitable place of production of chrysanthemum flowers is on the highlands of northern Thailand where the temperature during the summer is around 16-20°C. The same principle of flower forcing is applied, but in this case the daylength in northern Thailand may be lower than the critical value for certain varieties. As practiced in Doi Phu Kha of Nan Province, black cloth is used to completely cover the plant house from 16.00 to 08.00 hrs. for 30 days after the cuttings have been exposed to long day conditions (supplemented with artificial light).

Jasmine

Flowers are available thoughout the year, but very profuse during the rainy season and scarce during the winter. To produce jasmine for winter-season harvest, the following operations are recommended:

- One month before the planned date of harvest, stop watering for 2 - 3 days until the plants show sign of wilting.
- Prune the plant to a round shape so that blooming will emerge from mature branches when induced in stage (iii). In this way the blooms will be of large size and healthy as they receive full sunlight.
- Apply balanced fertiliser (e.g. 15-15-15) at the rate of 30 g/plant and water heavily. Keep watering normally

everyday. Flower buds will emerge within 10 days and blooming will occur within 25-30 days after pruning.

Amaryllis

Flowers are available all the year round, although depends on the availability of bulbs which are available from the Netherlands in the summer (June onwards). It takes about one month after the bulbs are planted to have full bloom ready for the harvest of cut flowers.

It is recommended to keep the bulbs in a refrigerator (4°C) for at least two months. The bulbs must be planted one month prior to the required date of flowering.

Gladiolus

Flowers can be available all the year round, provided the weather is optimum (requires cool climate). Blooms 90 - 100 days after planting.

In cool climate preheat corms before planting for 2 weeks at 27-32°C. This will force such corms to flower early. In warm climate soak in GA3 solution (10 - 25 ppm) before planting. This will accelerate flower by hastening differentiation of floral primordia. Growth retardants (e.g. CCC) promote initiation of floral primordia (by reducing endogenous GA3 level, or counterbalancing its inhibitory effect on floral initiation.

Roses

Flowers are available round the year but more profuse in the cool season.

For blooming during Christmas - New Year, and Valentine Day, the following procedure is suggested:

— Cut the branches in November. This will stimulate flowering during Christmas - New Year. Flowering will take place 43 days after this pruning. Cutting the flowers on 23 December will further stimulate flower bud initiation to bloom on 10 February. Flowering will take place 49 days after this cutting.

— Pinch (i.e. cut only the tip off) instead of pruning as in (i) above, on 10 November for 23 December harvest. This operation automatically stimulates flower initiation for 10 February harvest.

The following operations should be followed:

— Do not pinch healthy flower buds after 10 November as blooming will be due in 30 days. Allow them to bloom for 10 December harvest.

— Do not pinch flower buds which are still very small.

— Pinch all other branches, including those flower buds which are not healthy, deformed, etc. Do not pinch too far down to healthy leaves, unless they are water sprouts which should be cut off from the base.

4

Methods of Growing Annual Flowers

The term "annual" describes plants that go through their entire life cycle-from germination to seed production-in a single year. Because they last only one growing season, annuals are versatile members of the garden. In addition to offering a diverse palette of colour on their own you can also:

— Plant annuals among perennials or shrubs in a new bed where you need to fill spaces between small, developing plants.

— Use annuals to add accent in planters, boxes and urns (next year you can pick a new pallet!).

— Plant them over fading spring flowering bulbs. Annuals add colour in summer without interfering with next spring's flowers.

— Grow annuals in a garden for summer cutting or for drying to make winter arrangements.

— Create a temporary low-growing hedge or border of annual plants along a foundation.

As you choose annuals to plant, think about colour, form and texture. For example, you can choose lacey Nigella for a dainty effect or bright, coarse coneflowers for a bolder presentation.

Although most annuals bloom all season, there are some that offer only seasonal interest. The sweet pea and other cool-weather annuals quit blooming in the heat of summer. Coneflowers begin their show in mid to late summer. Pansies can be planted in fall to provide early spring blooms.

Most flowering annuals grow best in full sun and welldrained soil. Sun-loving plants that are grown in shade tend to be spindly and produce fewer flowers. Likewise, shade-loving annuals tend to get leaf scorch and flower poorly when grown in too much sun.

Few annuals will thrive when planted in wet, heavy soil. Before you plant, check to see how well the soil drains. Dig a hole 10 inches deep and fill it with water. If the water drains quickly, you have a droughty soil. Adding organic matter will aid water and nutrient retention. If the water drains after an hour or two, you have a welldrained soil. If the water stands overnight, you have poorly drained soil and may need to consider constructing raised beds.

If you're working ground that has remained unplanted for at least a year, add organic matter, such as peat moss, compost, or well-rotted manure the previous fall. For established planting beds, add organic matter several weeks before planting in spring. Spade or till the top six to eight inches of soil, thoroughly incorporating the organic materials.

To start annuals from seed directly in the garden, level the bed and rake it smooth after tilling. Remove all stones, clods, and old plant refuse. To plant annuals into a bulb or shrub bed, cultivate only lightly. You can add a little organic matter to help annuals become established, but excessive cultivation will damage the roots of established plants.

If you plan to grow your annuals in containers, use new potting mix each year. Try a mix of garden loam, sphagnum peat moss and vermiculite or perlite (1:1:1 by volume) or use commercially available mixes. Containers should be large enough to provide adequate space for root growth and must have drainage holes in the bottom to allow excess water to escape.

Add fertiliser to soil as you cultivate and incorporate organic matter. Use one to two pounds of 5-10-5 or similar analysis, fertiliser per 100 square feet of bed area. If the soil fertility is low, you may need to fertilise again during the growing season. Usually, ½ -1 pound of 5-10-5 per 100 square feet every month to six weeks will suffice. For home gardeners, remember that one pound of dry fertiliser is about equal to 2 cups or 96 teaspoons. Therefore, use 1-2 cups of dry fertiliser per 100 square feet, or about 1 teaspoon per square foot. Do not let dry or concentrated liquid fertiliser touch the foliage or flowers. Water thoroughly after applying fertiliser.

Seeds of a few plants, including babies-breath and gaillardia, can be sown as soon as you can work the soil in early spring. For most other annual flowers, wait until soil warms above 60°F before sowing outdoors. Most annuals can be started from seed indoors or in hotbeds or cold frames outdoors to get a head start on the growing season. You want to time the sowing of seeds so that plants will be ready for transplanting to the garden when the appropriate weather arrives. A few annuals, such as poppies and sweet pea, don't transplant well and should be sown in their permanent location.

Purchase fresh seed from local garden centers or reputable mail-order catalogs. Saving seeds from oldfashioned varieties can work well, but many of today's annuals are F1 hybrids. Seeds from these plants often grow into plants very different from their parents in colour, size, or habit of growth.

Smooth the seedbed, then make a furrow for the recommended sowing depth with a rounded stick or hoe handle. Place seeds in the furrow and cover them with a fine layer of soil or light mulch such as vermiculite. Water gently to moisten the furrow trying not to disturb the seeds. Good soil preparation will help promote speedy seed germination. However, in heavy soil, it can be helpful to cover the seed with fine vermiculite or finished compost to prevent a hard soil crust from interfering with seedling emergence.

Thinning

When the seedlings have developed two true leaves, thin them by pinching out excess plants or transplant them to another location. To allow each individual plant to receive enough light, water, nutrients, and air for proper growth and development.

Transplanting

Many homeowners find it more convenient to buy started seedlings rather than start their own. Most annuals transplant easily and often are already blooming when you plant them, though your selection will be more limited than if you started from seed. Choose vigorous looking plants with healthy looking leaves.

Most annuals should be transplanted outdoors after the danger of spring frost has passed: May 1, southern Indiana; May 15, central; May 20-30, northern.

If you've raised your own seedlings in a greenhouse or in some other controlled environment, it is helpful to harden them off before planting outside. The goal is to gradually increase exposure to wind, sun, and drought. Put the seedlings outside next to the house, and bring them inside at night at first. After a few days, you can leave them out at night but protect them if the temperature dips close to freezing. After a week or so they'll be ready to plant out in the open. To remove seedlings from trays or flats, slice downward through the soil between the plants with a knife. To remove seedlings from fibre, plastic, or clay pots carefully lift out each plant with its intact block of soil.

To transplant seedlings in peat pots, tear off the lip of the pot that would otherwise extend above the soil surface as a safeguard, or the lip may absorb water away from tender seedling roots (wicking). Peat moss can be difficult to wet once it becomes dry, forming an impenetrable barrier between the roots and the surrounding soil. To ensure root penetration, poke holes or remove the bottom of the pot. You can plant the sides of the pot with the seedling.

Set the plants at the same depth they were growing in the original container and at their recommended spacing. Press the soil firmly around the plant. To help give the root system a boost, water with a dilute, high phosphorus fertiliser solution, such as one tablespoon of high phosphate fertiliser (10-52-17 or similar analysis) in 1 gallon of water.

Mulching Flower Beds

Mulching flower beds not only makes them look better, it conserves moisture, reduces weed growth, and can help moderate soil temperature. Common mulches include straw, wood chips or shreds, ground or crushed corn cobs, pine needles, and black plastic. Apply 2-3 inches of mulch around the plants after soil has had a chance to warm.

Black plastic mulch will warm the soil and is particularly effective for heat-loving plants such as petunias, marigolds, and zinnias.

Pinching

Most annuals benefit from pinching, resulting in more flowers during the growing season, even though it may remove blooms initially. The first pinch should remove the top inch or two from the growing tip, leaving 3 or 4 leaves. Greenhouse-grown transplants have usually been pinched by the grower so check the plants at purchase time. Many annuals that bloom all summer will benefit from additional pinching throughout the summer to keep them compact and full of blossoms.

Cultivation

Lightly cultivate soil early in the season to break the crust that forms on the surface. After the plants are growing vigorously, stop cultivating. When you cultivate large plants you may damage some of the shallow feeder roots. Cut weeds off with a shaving stroke at the soil surface instead of pulling them. Pulling weeds usually brings other weed seeds to the surface, where they will germinate. Mulching will eliminate the need for cultivation.

Watering

Generally, plants need a single watering of 1 inch of water each week. Watering with a sprinkler or soaker hose works better than with a hand-held hose since it can take 1-3 hours to provide sufficient water. Allow the water to penetrate at least 6-8 inches into the soil. Soaker hoses are most efficient, since they apply the water slowly and directly to the soil, reducing water loss to runoff and evaporation.

Flower Removal (Deadheading)

Annuals are prolific bloomers on their own but removing old, faded blooms every 5-7 days will help encourage additional blossoms. This is called deadheading. When you leave spent flowers untouched, the plant puts its energy into developing seeds rather than into producing more flowers. In some cases, the fruits or seeds are themselves ornamental and you'll want to let them develop. Ageratum, calendula, cosmos, marigold, pansy, rudbeckia, scabiosa, and zinnia respond especially well to deadheading.

5

Home Gardening

There are so many joys in growing house plants. Thousands of people grow plants and more start everyday. There are lots of advantages of home gardening like fresh air all day long to breath, nice aroma from the flowers, home grown vegetables etc.

African home gardens vary, depending on agro-ecological, climatic and cultural factors. In the humid and subhumid areas of central and West Africa, a home garden often includes a permanent agricultural plot or forest garden that contains perennial and annual crops. In some cases, the garden surrounds the homestead and paths lead to other field systems and production units devoted to annual crops for the market and home consumption. In other cases, the garden is located at a distance from the homestead, but close to a water source.

The biological diversity and complexity of home gardens decline with the transition from humid areas, where annual rainfall exceeds 1 100 mm, to semi-arid and arid regions where annual rainfalls range from 500 to 900 mm, to 250 to 500 mm in the Sahelian zones. Insufficient water is a major constraint to successful gardening in dry areas, yet even in these areas, gardening is possible. Some crops can be grown in the dry periods through appropriate soil management, and with cheap and effective solutions for harvesting and storing water.

The home garden visit will help participating field workers gain an understanding of:

— the home garden as the most direct way of providing daily food throughout the year, particularly vegetables and fruits, and its function as a safety net during the lean season;

— the average size of local home gardens, the variety of crops grown and seasonal variations;

— the contribution that a well-developed, as opposed to a not-well-developed, home garden can make to family food security.

Household Food Security and Nutritional Status

Having sufficient good-quality food available for household consumption depends on several factors. Household production is one way to ensure that there is food within the home at all times, but producing enough of a variety of foods in the home garden and in the fields requires access to adequate resources, including land, water, seeds, tools, knowledge, skills and labour. Roads and transport to markets are necessary for buying and selling food and other essentials. Household members also need access to other commercial and government services, and off-farm employment during periods of low agricultural labour demand.

In rural areas with limited income-earning opportunities, the ability to produce most foods in the home garden and on the farm, without depending on market purchases, means a better guarantee of household food security. Household food security depends on a regular and sustainable supply of food throughout the year. In many rural areas, however, households often face food shortages because crop production is seasonal and at times inadequate.

Household Food Security, Care and Health

Household food security

It is defined as "access by all people at all times to the food needed for a healthy life". Households can obtain food supplies either from their own food production or from food purchases, but more often it is through a combination of both. Some factors that help assure that communities have enough of a variety of foods at the household level are access to sufficient water, fertile land, seeds, planting materials, agricultural implements, extension advice, credit, good storage and a sufficient number of family members who are healthy and strong enough to work on the farm and undertake off-farm employment.

Many rural communities, however, do not have year-round access to adequate amounts of either fresh or processed staple foods, and their access to fresh vegetables and fruits tends to be seasonal. Frequently, households sell too much food either because they need cash, or facilities for storing and preserving foods for longer periods are poorly developed or non-existent.

Households can use several strategies for ensuring continuous access to a variety of nutritious foods. These include the year-round production of a variety of home garden foods (given that some water is available) and the preservation, processing and adequate storage of foods.

Households can thus insure themselves against seasonal shortfalls, but it requires planning ahead and thinking about what is the most effective use of available resources so as to prevent hungry periods. Appropriate technology options are available to extend garden production and enable households to process and store perishable staples, legumes, vegetables and fruits, in order to extend their availability and enhance their marketing potential.

Care

Household food security is a precondition for adequate dietary intake, but it does not always translate into good nutritional status. The provision of adequate care especially for small children and other nutritionally vulnerable household members is crucial to achieving good nutritional status.

Care includes such things as having knowledge of the differing food and nutritional needs of household members. Such knowledge helps the caretaker, usually the mother, make the right decisions on what foods and how much to prepare and serve during meal times to the children and adults. Apart from having nutritional knowledge and skills, the caretaker must also have adequate time for child care, food preparation and child feeding.

In many African cultures, women are the main household food producers and caregivers. They also have many other responsibilities, such as fetching water and fuelwood, and storing, processing and marketing food. During peak seasons of agricultural activity, the danger is particularly high that children's needs for care, proper food and frequent feeding will not be met adequately. Despite the many constraints, there are many options to enable households to better care for their vulnerable members. These include the use of available technology that relieves demands on women's time, such as providing better access to water.

Other options are more people-centred and require the community or individuals to change established traditions or behavioural patterns. This means that all community members, adults and young people, should become better educated as to the advantages of good nutrition and health, as well as the many possibilities that exist for community and individual action, such as establishing day care centres or encouraging the active participation of fathers in child care.

Health

In addition to household food security and care, people have to be healthy to get the full nutritional benefits from food, otherwise some of the nutrients will be wasted. This means the nutrients are either passed through the digestive system unabsorbed, as in the case of someone suffering from diarrhoea, or consumed by parasites such as worms. This is why households must work to prevent communicable diseases and parasitic infections by practising good environmental and personal hygiene, by drinking safe water and keeping food safe and clean. If household members become sick, they should visit the health centre or seek advice from the village health worker. If not treated, infections that lead to poor utilisation of available food and nutrients in the body can result in malnutrition.

Variety of Food and Biological Diversity

Planting small amounts of a wide variety of vegetables is a sound environmental strategy, as it conserves biological diversity and contributes to food security and nutrition. Women farmers are particularly interested in maintaining crop diversity, since they are often the ones who use these genetic resources to develop new varieties according to their families' changing needs and preferences. Growing a diversity of crops reduces the risk of a total crop failure as a result of disease. It also helps households vary their diets, prepare a broader range of dishes and enhance the nutritional quality of their meals.

During the rainy season, households have access to several traditional leafy vegetables that grow as either uncultivated or semi-cultivated vegetables on the farm, but these varieties may not be readily available during the dry season. Their inclusion, therefore, in the home garden cropping pattern, particularly during the dry season, supplements the dry-season vegetable supply and safeguards against crop failure, since traditional crops are well adapted to the local environment. To ensure the availability of seeds from such vegetables during the dry season, farmers must take care to collect their seeds at the end of the rains, store them properly and plant them in the home garden under irrigation.

Contribution of the Home Garden

Improved Crop Production in the Home Garden

The improvement and maintenance of soil fertility are major challenges facing most home gardeners in Africa. They must, therefore, have access to the appropriate information and technical advice on:

— locally available organic matter that decomposes within a few weeks to make compost;

— crop-rotation and intercropping practices that minimise the depletion of plant nutrients in the soil;

— types of crops that can be planted as green manure;

— integration of crop and livestock production in a manner that creates animal manure best suited to the types of crops grown by the local community;

— chemical fertilisers (where affordable) that are essential for complementing compost and animal manure.

Home garden managers must have access to information on the fertiliser types best suited to specific home garden crops and the appropriate times for fertiliser application.

The participants should help farmers choose among the different technology and home garden management options. Where appropriate, participants can demonstrate soil management, soil fertility improvements and other relevant technologies, either on demonstration plots or, preferably, on the farmers' own land. The field workers should exploit all available means of disseminating information, including radio, television, newspapers, songs and local theatre.

Selection of the Technologies and Plant Varieties

In addition to soil management and the improvement of soil fertility, a wide range of technologies exists for improving home garden production and productivity. The area's agro-ecological and climatic characteristics and its altitude largely determine the types of crops to plant and the seasons during which those crops grow well. Technology options, therefore, should be directly related to the altitude and agro-ecological and climatic features of a given area. Farmers, therefore, require information on:

— the most appropriate water management and water-harvesting techniques;

— the crop varieties best suited to their agro-ecological zones and the seasons during which those crops grow well;

— low-cost and sustainable pest and weed management techniques.

Water Management and Water Harvesting.

The annual rainfall pattern and access to reliable watering facilities often determine the physical location of a home garden and the types of beds a farmer may construct. Farmers must have access to appropriate information in order to know whether to construct sunken, flat or raised beds or mounds, depending upon which crops and which times of the year.

Regions with steady rainfall throughout the year have greater potential for producing a steady, year-round supply of fruits and vegetables. On the other hand, communities in areas with distinct wet and dry seasons must have a variety of strategies for ensuring an adequate, year-round supply of fruits and vegetables. Such strategies can include developing two household garden plots, one around the homestead and exploitable mainly during the rains (i.e. rainfed), the other one away from home in a wet lowland or swampy place. For the household garden plot to produce vegetables and other crops throughout the dry season, the use of unconventional water sources, such as household wastewater (from washing dishes or bathing) might be an option.

In addition, communities in areas with inadequate access to dry-season irrigation water can secure a communal water point, by constructing small dams or hand-dug wells, paid for by the pooling of community resources. Next to this water source, participating households can plant a "community garden" in which each household has its individual plot but shares the water and fence with the other participating community members.

Assisted by the field workers, community members can jointly decide on the community garden's planting patterns and times for specific food crops. Seed and other gardening inputs can be purchased with help from a community fund or with savings accumulated through group savings schemes. Such community gardens provide an excellent contact point for the delivery of technical information by the agricultural, nutritional or health extension workers in the area.

Crops and crop Varieties

While the integration of traditional and local vegetables into the home garden cropping pattern should be encouraged, the low yields of such crops inhibit productivity increases. However, home garden managers can be taught to maintain balance between the traditional and the improved, or introduced, varieties.

Encouraging the cultivation of a broad variety of food crops is a greater challenge in the semi-arid and arid regions and under low-temperature conditions (e.g. in highland areas) than it is in the humid areas. Efforts to extend the production of fruits and vegetables into the drier and colder periods of the year have to be based on the selection of hardier crop species and varieties, together with the application of appropriate cultural practices such as:

— raising early seedlings for transplanting;
— using windbreaks;
— adding mulch and organic matter to improve the soil's water-holding capacity;
— protecting crops from low temperatures with the use of plastic sheeting or straw;
— using all available water supplies to keep crops growing during dry periods.

Low-cost Pest Management

The control of plant pests and insects, particularly during dry-season home gardening, is another challenge of rural households. Farmers must have access to information on developing good home garden hygiene. In addition, they must be aware of environmentally friendly methods of controlling pests, for example:

— intercropping food crops with chilli, garlic, basil and other plants that repel certain pests;

— using natural pesticides (e.g. wood ash, tobacco, neem, papaya leaves) or easily made preparations such as soap or vegetable oil emulsions or starchy suspensions;

— removing and burning diseased plants;

— intercropping and rotating crops of different family types to check the spread of diseases and pests specific to a particular crop type.

Carefully Selected Planting Pattern

When planning home garden improvements for a continuous supply of fruits and vegetables, it is important to take into account untapped resources and potentials as well as household food needs and preferences. Factors such as the length of a crop's growing period (from planting to maturity) and the length of time during which a household may harvest edible fruits or vegetables from a mature plant before it must be replaced should be considered when planning a home garden planting pattern. In addition to taking into account basic crop-rotation requirements and desired intercropping aspects, home garden managers must stagger the planting of different types of crop.

To reduce gaps in the year-round supply of fresh fruits to the household, for example, the home garden manager must select and plant fruit-tree types with fruits that mature at different times during the year. These can be mixed with fruit-trees that bear fruit throughout the year (e.g. bananas).

Leafy vegetables and legumes must be planted in a manner that provides the household with access to some varieties of green leafy vegetables and fresh or dry legumes throughout the year. Vegetables that grow well all year round (e.g. some varieties of leafy vegetables) can form the core of the home garden, and seasonal varieties can be planted when appropriate.

Information on the types of fruits, vegetables and legumes that give the best yields at specific times of the year is crucial, because it enables the home garden manager to organise planting patterns for optimum yields and nutritional benefits. Home garden managers should also be sensitive to the planting patterns of other local farmers and, thus, can avoid flooding the market with one or two food commodities, should their household produce surplus for sale.

Planting a mixture of annual and perennial crops (e.g. cassava and fruit-trees) and adopting multiple and multistorey cropping techniques increase the year-round availability of food to the household. Sweet potato and cowpea can be useful cover crops. Small numbers of different crops and fruit-trees, and fresh vegetables in particular, planted and harvested consecutively, provide a constant supply of home garden foods.

If adequate quantities and varieties of home garden foods are to be produced, particularly in the wet lowlands where year-round food production may be possible, intensive crop production techniques must be considered.

The use of better cultural practices (i.e. soil fertility and water management, optimum planting times and plant spacing, proper crop rotations, weed control and integrated pest management) can increase the productivity of individual crops. As more intensive crop production practices are adopted, better post-harvest handling techniques (i.e. processing, preservation and storage) must be developed in order for farmers to gain maximum benefits from the increased garden output.

Finally, households should protect their home gardens from roaming animals, especially during the dry season. Using live fencing for this purpose will provide further benefits to the household (e.g. fuelwood, trellising and medicines). The agricultural extension worker can help families and communities identify the crops that are suited to local conditions and that will supply the nutrients needed for the family to achieve optimum nutritional balance year round. Since women are usually responsible for home gardens, their full participation in planning and implementing improvements is essential.

Improving Nutrition through Home Gardening

Individual consultations with households are valuable but cannot give families all the knowledge and skills they need to make their own decisions and then act on them. The community, as a group or in several groups, must take actions that make people aware of the importance and urgency of improved food production and better nutrition, and then be assisted in implementing those actions. Information and practical training in home garden technologies and nutrition are needed to achieve these goals. This is where agricultural extension workers and other field workers can provide valuable support.

Community Action

The participants learned to recognise the importance of household food security (i.e. having access to enough nutritious foods year round) and the home garden's potential contribution to food security. They applied their knowledge both to case studies and to individual real-life situations.

The next thing to consider is how the entire community can improve its household food supply. An initial step towards developing the different components of a community plan of action is to list the solutions that community members consider feasible. Households may be able to implement some of these solutions immediately, while other solutions may require them first to mobilise resources and acquire practical skills.

There also may be a set of problems that community members do not necessarily perceive as problems because they lack the knowledge (e.g. the causal relationship between parasitic diseases and malnutrition). Participants should identify these types of knowledge gaps.

The strategy for action to be adopted must therefore involve:

— raising community and household members' awareness of the importance and urgency of improving a home garden's production for better nutrition and extra income;

— encouraging and supporting them, through practical skills training, in taking immediate action using existing resources.

With the list of activities established, community members can discuss them and then decide jointly how, by whom, when and where those activities should be undertaken, and what inputs and technical assistance may be required to implement them. Knowledge of the different elements and stages of planning enables field workers to facilitate this step-by-step process.

Action to Promote Improved Nutrition

Any project or programme involving a number of households will need careful planning. One of the most effective ways of planning an action programme is to work systematically through the elements of project formulation:

— rationale;

— goal;

— objective;

— strategy;

— outputs and targets;

— activities;

— inputs;

— timetable and work plan;

— monitoring;

— evaluation.

Planning a project is like planning a journey. Before you embark, you need to decide:

— why you are going (rationale and goal);

— where you want to go (objective);

— how you propose to get there (strategy);

— what the immediate results of your trip will be (output);

— when you expect to arrive (target);

— what preparations you need to make and how you will travel (activities);

— what you will need, e.g. documents, vehicles, petrol (inputs);

— how long it will take (timetable);

— what progress is being made along the way (monitoring);

— has the objective been achieved (evaluation).

A methodical approach makes it possible to:

— determine how realistic the project is and whether it can be implemented with locally available resources and expertise;

— anticipate problems;

— check on the progress of the work (monitoring) and on whether the project has achieved its aims (evaluation).

Soil Management

Soil is formed from decomposing rock. Depending on the type of rock, different soils have developed over the centuries and are still developing. The structure and characteristics of soils and their capacity to support plant growth differ according to the size of the soil particles and the content of organic and inorganic matter.

If the soil is kept in good condition - protected and "fed" with the nutrients that the plants require - the home garden can be cultivated year round and year after year.

Preventing Soil Erosion

The first step in soil management is preventing the loss, or erosion, of soil. Topsoil is particularly vulnerable to erosion if not protected by plants or mulch or by other measures. The soil that remains after the loss of topsoil is usually less productive, which results in lower production from home garden crops. The challenge is to protect home garden soil while using the land for food production and other non-food activities.

Soil erosion is caused mainly by wind and water but also by incorrect cultivation practices. Rain and wind dislodge and then carry away soil particles. Where the soil is bare or the vegetation poor, rainwater does not seep into the soil; instead it runs off and carries with it loose topsoil. Sloping land and light soils with low organic matter content are both prone to erosion. Once eroded, the soil is lost forever.

Soil erosion is a problem in regions with little vegetation, particularly in the semi-arid and arid zones of Africa. In the humid tropics, erosion was not considered a problem when the land was in its natural state, because the variety of native plants kept the soils covered at all times. Now, people are clearing more land for agricultural purposes, and the situation has changed. Heavy rains coupled with poor soil management of cultivated areas are now common causes of soil erosion in the humid areas.

Preventing Water Erosion

There are three common forms of water erosion.

— *Sheet erosion:* a thin top layer of soil is removed from the soil by the impact of rain. With sheet erosion, small heaps of loose material (e.g. grass) amass between fine lines of sand after a rainstorm. This erosion takes place across a whole garden or field.

— *Rill erosion:* water flows over minor depressions on the land's surface and cuts small channels into the soil.

The erosion takes place along the length of these channels.

— *Gully erosion:* a gully forms along natural depressions on the soil's surface or on slopes. The head of a gully moves up the slope in the opposite direction of the flow of water. Gullies are symptoms of severe erosion.

Preventing Wind Erosion

This occurs mostly on light soils and bare land. High winds cause severe damage. Wind erosion is a common problem in dry and semi-arid areas, as well as in areas that get seasonal rains. Unlike water which only erodes on slopes, wind can remove soil from flat land as well as from sloping land; it can also transport the soil particles through the air and deposit them far away. Soils vulnerable to wind erosion are dry, loose, light soils with little or no vegetative cover.

Ploughing up and down a slope causes soil erosion. To prevent the loss of home garden soils, certain measures must be taken. These include:

— clearing only the land to be cultivated;

— planting along a contour and using grassed channels;

— establishing windbreaks and bench terraces;

— ploughing along a contour;

— planting cover crops and mulching.

When clearing land for cultivation, the beneficial effects of certain trees and plants should be considered. Some trees should be left, since they may supply food, medicine, shade or, when they shed their leaves, organic matter.

Feeding the Soil with Organic Matter

One of the main goals of home garden development is to

make the garden's soil fertile and well structured, so a wide range of useful crops can grow and produce well. In order to grow, plants require nutrients that are present in organic matter, such as nitrogen, calcium and phosphorus, as well as minerals and trace elements.

If the natural fertility or structure of the soil is poor, it must be continuously "fed" with organic matter, such as leaves and manure, in order to improve its productivity and water-holding capacity. As organic matter decomposes, it becomes food for plants. It also improves soil structure by loosening heavy clay and binding sandy soil. Feeding the soil with organic matter is especially important in the early years of home garden development. Organic matter (i.e. waste from plants and livestock) can be collected and buried in the soil, where it will decompose. The organic matter also can be used to make compost, which can be applied to the soil to enrich its fertility.

The roots of legumes contain nitrogen-fixing bacteria. Therefore, intercropping or rotating legumes with other crops helps maintain or improve the nitrogen content of the soil, and this enhances the growth of other plants. Healthy plants yield more and are better protected from insects and disease. The application of organic matter, such as compost, animal manure, green manure and soil from anthills, improves soil structure and adds nutrients to the soil.

The ideal way to protect and feed the soil is to apply organic matter or compost regularly and to keep the soil covered with plants. A multilayer cropping system in which a mixture of trees and other plants with different maturity times are grown together will protect the soil and recycle nutrients. Leguminous plants such as cowpeas, groundnuts and beans are particularly useful in providing continuous nutrients for garden crops.

Plant Nutrients

Nitrogen is important for the growth of the stems and leaves of home garden plants. It also contributes to the development of chlorophyll (a plant's green pigment). Chlorophyll is needed during photosynthesis; a process by which a plant makes carbohydrates (starch and sugars) from water and the carbon dioxide in the air. Pale green or yellow leaves, poor growth, leaf fall and pest problems are indications of nitrogen deficiency in plants.

The application of green manure and compost can improve the nitrogen content of soil. Ground-up horns and bonemeal from slaughtered animals are rich in nitrogen and also contain phosphorus. They can be applied to the soil, but this form of natural fertiliser reacts slowly in soil. Shredded Ricinus (castor oil) plants are another source of nitrogen and phosphorus. They are particularly useful for crops with high nitrogen requirements (e.g. young plants of maize or sorghum).

Phosphorus reacts slowly and is important for root formation and the formation of seeds and fruit, as well as for the development of a plant's drought tolerance. Phosphorus deficiency symptoms are stunted growth and poor flower and shoot development. Rock phosphate, wood ash and animal bonemeal and manure supply phosphorus. Chicken manure is a good source of phosphorus, nitrogen and potassium, but it is quite strong and must be composted or diluted before application.

Potassium contributes to the development of roots, fruits and seeds and improves the storability of crops. Poor plant growth and uneven ripening of fruits and seeds are symptoms of potassium deficiency. Besides chicken manure, compost, ash and banana leaves also improve the potassium content of soil.

Soils and Plants

Plants and soil are symbiotic, i.e. they are interdependent and require a good balance of give and take. Soil nutrients are absorbed by plants and then recycled back into the soil in the form of plant residue. When more nutrients are removed from the soil than can be replenished by the application of organic and inorganic matter, the soil gradually becomes poorer and unable to support the growth of healthy plants.

Plant residues must decompose in order to release plant nutrients. The microorganisms, worms, fungi and bacteria found in soil are responsible for decomposing organic matter and producing humus (the fertile part of the soil). Most of the humus is found in topsoil. Subsoils contain less organic matter and are therefore less fertile. Some soil nutrients react slowly and are stable (e.g. phosphorus). Others are absorbed quickly and used up by plants (e.g. nitrogen). These must be replaced continually in order to keep a good balance between supply and demand.

Excessive use of chemical fertilisers is discouraged. Home garden managers are instead encouraged to use natural soil additives, such as different types of manure, in order to maintain the natural balance between soil and plants. Natural fertilisers such as green manure, compost and animal manure provide plant nutrients but also improve the structure of the soil.

Allowing soil to rest and restore its natural fertility is called fallowing. Fallowing is practical, however, only if land is available. When land is left fallow, it is important to prevent weeds from seeding on it. In this case, the use of green manure or cover crops during fallowing is recommended. In addition, crop rotation of deep- and shallow-rooted crops, and crops with different nutrient requirements, is also recommended.

Water Management

Soil and water management are closely interrelated. Although there are great differences in the water requirements of different plants, no plant growth is possible without a certain amount of water. Out of necessity, however, a number of plants have developed great drought tolerance, and some even drought resistance.

Plants can contain up to 90 percent water. The water is absorbed mainly through the root system of the plant. With the water, plant nutrients are absorbed. Healthy roots need air (aeration) for development. Excess water in the soil prevents air from penetrating and damages a plant's roots. Water management is therefore extremely important in regions with good water resources as well as in those where water is scarce. The water-holding capacity of soil varies according to soil type. Soil with a high content of organic matter has better aeration, better structure and better water-holding capacity.

Heavy, sticky soils are too dense to allow air in and water out, so roots cannot breathe and plants can have growth problems. When this kind of soil dries out, it sets like cement, and water takes a long time to soak into it. On the other hand, sandy, coarse-grained soils are too loose to hold water before it drains away. In this kind of soil, without a regular external water supply, a plant's roots cannot find enough water for growth. Regular application of organic matter will improve the ability of both these kinds of soil to hold and release enough water and air.

High Rainfall Areas

In high rainfall areas or wetlands, water management involves limiting the damage from temporary excess water. Drainage is therefore the most important factor. On sloping land, cut-off drains made along contour lines lead the water away. The gradient should not exceed 0.4 percent (i.e. a difference in elevation of 10 cm in 25 m) to slow the flow of water and avoid erosion. Water can be collected for later use in a hand-dug pit at the end of a cut-off drain.

A good way to manage excess water is to put plants that can tolerate more soil moisture near the water source (wetlands, the banks of a stream or a flooded area). Less water-tolerant plants can be planted on mounds or raised beds. If properly managed, some wetlands (which are known as dambos in Zambia, mapani in Zimbabwe, mbugas in the United Republic of Tanzania and marais in Rwanda) can be used to grow crops year round. Excess drainage should be avoided, especially on organic soils.

Low-lying wetlands are subject to seasonal or permanent flooding. This may be the result of surface runoff or groundwater seepage from a catchment area over an impermeable soil layer towards low-lying areas. In such areas, organic matter decomposes slowly, forming the black, mulch- and humus-rich topsoil that is associated with high soil fertility. Wetlands are important for smallholder food production for a large part of the year, especially during the dry season. In East Africa, southern Africa and some northern parts of

West Africa, home gardens are mainly found in such wetlands. Since the wetlands can remain wet for a considerable part of the year, most crops are grown on raised beds or ridges (also referred to as mounds) to reduce waterlogging and facilitate drainage. Ridges will keep excess water away from the plants and lead it directly to the plant's roots. When the water table falls, water is channelled into pits and scooped with tins on to the ridges. If the water table sinks farther, plants are grown in flat beds and, eventually, at the peak of the dry season, sunken beds. Also plants can be established on selected parts (top, side or bottom) of the ridge according to their water requirements and waterlogging tolerance.

Canals and trenches drain water from rain-flooded areas. Plants that require frequent watering, such as rice, sugar cane, taro and water spinach, can grow in them. Other plants, such as cassava, yam and leafy vegetables, can be grown on raised beds. Wetlands are delicate environments and should be used with great care. Both soil erosion and excessive drainage damage wetlands.

Dry Climate or During the Dry Season

Maintaining soil moisture and collecting water from different sources is more difficult and labour intensive than keeping excess water off crops. Water management in dry areas requires efficient use of available water resources through the prevention of runoff and the reduction of evapo-transpiration, i.e. the prevention of loss of water through evaporation from the soil, and from the leaves of plants.

Under dry conditions, plants should be placed to make optimum use of available moisture. It is, of course, most important to choose plants with drought tolerance and low water requirements. Deep-rooted plants, such as African eggplant, can usually survive periods of limited water availability better than shallow-rooted species. Once established, permanent crops, such as many trees and shrubs, will withstand reduced water conditions as well.

It is also important to make the best use of rainfall and make moisture available to plants for as long as possible. During the rainy season, basin-like or sunken beds are often used to keep a maximum amount of water accessible to plants and to prevent runoff of surface water. Microbasins have also been in use for a long time and are well suited to sloping land as well. They are built by placing a ring or semicircle of stones, soil or other material around individual plants, in particular tree crops. It is important not to heap the soil too high on the plant's stem, as this might damage the bark. Instead, water should be able to collect in a ring above the root line, some distance from the stem.

If plants suffer from direct sunlight and heat, water is lost through transpiration. This can be prevented or reduced by creating natural windbreaks and providing shade. Mulching will also reduce evaporation of moisture from the soil.

Weeds should be removed because they compete with cultivated plants for moisture. To reduce crop failure, short-term crops can be grown near the water source. High organic matter content in soil helps retain the soil's moisture. Compost and green manure help increase the organic matter content of soil. Mulching helps hold water in the soil by keeping the soil surface from drying out and becoming too hot. Light-coloured mulches, such as straw, are particularly good, since they reflect sunlight and heat away from the soil.

Community Mobilisation

In the humid and subhumid parts of Africa, crops grow under rainfed agriculture for about seven months of the year, and the demand for irrigation water is relatively low. In addition, the water table is fairly high during the rainy season; therefore, the cost of securing a reliable water source for home gardening purposes is also relatively low. For households in the semi-arid regions, however, the situation is quite different. Individual households in these areas often do not have access to a reliable year-round water source.

In such areas, the need for a reliable water source can become an effective driving force for mobilising a community. This can be achieved in different ways. For example, households, in consultation with the relevant technical extension sectors of governmental or non-governmental agencies, can form groups and decide on the most appropriate water source to develop. Interested households can then pool their resources to establish a common water source, for example, by building a small dam, digging a well, or procuring water pumps to lift water from a nearby river.

Community members can form savings groups or associations to obtain matching grants from a community development or microproject fund for the purchase of water pumps. Investing in a regular supply of water has multiple benefits if water quality can be assured. Apart from providing water for year-round food production, it can also supply sufficient and clean drinking-water, and alleviate women's workload by reducing the time involved in fetching water from a distant water source.

The mobilisation of a community around a common water point can form a nucleus for complementary development activities. For example, the community water association can provide a means of organising or improving the marketing of home garden produce if there is surplus; make joint purchases of inputs (e.g. seeds, tools) and other essentials; or create an effective force for bargaining for better prices. After investing in securing a common water source, communities need to take an active role in ensuring its effective functioning through regular maintenance.

Weed and Pest Managemfnt

Problems of weeds, insects and animals reduce home garden production and cause frustration for the home garden manager. Effective crop protection requires an understanding of the types of pests and diseases that attack home garden crops and an awareness of the different means of controlling them. Often, the problem will disappear if the selected plants suit their location and if the soil and water are well managed.

Plant disease is an abnormal condition caused by small organisms usually too small to be seen, called bacteria, fungi, viruses, or nematodes, which are extremely small worms that attack roots. Injury to plants is also caused by pests, such as insects, mites or worms. Some plant diseases are easily spread. For example, whiteflies transport the viruses that cause mosaic disease from diseased cassava plants to healthy ones. Some parasitic weeds, such as striga and ororbanche, are considered pests because they cause heavy crop losses.

Weeds

Weeds are a major problem for food plants in many home gardens. They compete with food crops for nutrients, water, sunlight and space. Weed-infested crops grow poorly and, at times, die. Densely weeded areas can harbour rats, snakes and insect pests. Weeds tend to be a bigger problem in home gardens that have no mature trees. If not well managed, weeds can require labour that could otherwise be used for cultivating useful plants.

Grass weeds can be a big problem because they can invade a garden quickly. On the other hand, they can also help a garden by preventing soil erosion and producing material for mulching and compost, or for roof thatch.

The main ways to control weeds are to shade them from sunlight and deprive them of a place to grow. Three techniques for doing this are to:

— cut or dig weeds using a hoe or sickle before the flower releases seeds;

— cover the ground with 6 cm of mulch (weeds cut by hoe or sickle can be used as mulching material);

— plant quick-growing vine plants that will cover the ground (examples of such cover crops are pumpkin, cowpea and sweet potato).

For more permanent weed control, plant crops that permanently shade the soil's surface. The multilayer cropping system, in which plants of different heights are grown together, is the most effective system.

Pests and Diseases

Insect and disease problems tend to be seasonal, with the greatest problems occurring during the rains. In order to know when and where pest problems are likely to arise, it is important to learn the habits and life cycles of pests. Good home garden managers know as much about their plants' enemies as they know about their plants.

Insect pests. These can often be seen on plants, or in or on the soil. They damage plants mostly by chewing the leaves, roots, shoots and fruits, or by sucking sap out of the leaves, stems and fruits. Not all insects are pests. Bees facilitate the pollination of flowers. Other insects, called predators, feed on pests and reduce their population numbers.

Fungi. These attack all parts of plants. Symptoms of fungal infection are the appearance of a powdery substance on and under the leaves, rotten patches, spots on the leaves, fruit and stems, and wilting caused by affected roots. Fungi may be spread by rain or wind, or from one affected plant to another by cultivation tools.

Bacteria and viruses. These are not easily seen by the naked eye. They can cause rot in the root zones, oozing, distortion and a change of colour in the leaves and stems. Some of the symptoms are similar to those caused by fungi. Bacteria and viruses are spread through water and soil, and by affected plant materials. Unlike the rot caused by fungi, bacterial rot is usually characterised by a foul smell.

Some nematodes are soil-borne pests that attack roots, resulting in the abnormal growth of a plant's roots, stunting and generally poor growth. Affected plants often have knots and galls on their roots.

To Prevent and Manage Pest and Disease Problems

Weak plants suffer more from insects and disease than healthy plants. Good crop management, including proper attention to water, soil and weeds, will help reduce damage from insects and disease. To prevent and manage pests and diseases in their gardens, home garden managers must adhere to these general rules.

Choose crops according to the season and the location. If an annual plant does best in full sunshine and when planted at the beginning of the dry season, the home garden manager should ensure that it is planted under these conditions. A plant grown in the wrong season or the wrong location will be more vulnerable to attack.

Select plants that grow well in the local climate. Plants from other climates may not grow well in local climates. For example, mango produces well in hot lowland areas, where the dry season coincides with the flowering and fruiting, but if it is grown in an area that is humid year round, insects and disease will damage its flowers and its fruit harvest will be poor.

Rotate vegetable crops regularly. Rotate vegetable crops, planting them a few metres away from where they grew before, and plant a different crop in the original spot (e.g. plant legumes in a place where amaranth was previously grown). This will help prevent the build-up of disease infection in the soil.

Remove diseased leaves and other parts from plants. Reducing the amount of material that insects and disease can feed on will slow their spreading.

Sprinkle wood ash. Sprinkling wood ash around the base of young plants, and also lightly on new leaf growth, deters insects and fungi.

Grow plants that are known to repel certain insects. Chilli, lemon grass, basil and marigold are common plants that produce strong smells which repel certain insects.

Use home-made pest sprays. Sprays produced from neem fruits; tobacco, chillies or soap and water deter chewing and sucking insects.

There are many synthetic pesticides that can be extremely efficient in controlling pests and diseases. However, in an African home garden, they are rarely necessary or appropriate. With the great number of crops being grown simultaneously in a home garden, the use of chemicals to control pests for one crop can easily damage another. Lack of money to buy pesticides and lack of expertise and experience in applying them - many are harmful to human health when applied incorrectly - are some of the reasons for encouraging the use of good production practices, natural control methods and available natural pesticides. When chemical pesticides must be used they should be applied judiciously and with extreme caution.

Damage from Animals

Chickens often roam freely in home gardens. Although they catch insects and their manure fertilises the soil, they can also become pests when they attack and seriously damage plants in their search for food. Other domestic animals, such as cattle, sheep, pigs and goats can also be a threat to home gardens. Goats and pigs sometimes roam around villages and will eat many kinds of plants. Damage from animals can discourage home garden managers to the point of abandoning home food growing.

Wild pigs, where they are not hunted for food, can be persistent and often evade farmers and cross weak fences, but living fences can keep them out successfully. Good methods to deter animals from entering a home garden and protecting plants from damage by animals are:

- — creating a living fence of pineapple densely planted together with Gliricidia trees to keep out wild pigs;
- — planting cassava or placing sticks around the base of a plant so that chickens cannot dig up its roots or peck at it;
- — constructing a fence made from appropriate local materials;
- — keeping domestic animals in fenced-in areas.

Crop Management

There are many different farming methods suitable for the home garden. In each method, the home garden manager oversees the different parts of the system (i.e. soil, water, weeds and crops) to ensure that the garden can produce as much of a given crop as possible, now and in the future.

This means that the home garden manager must take into account the home garden's future production.

Important Aspects

Crop management depends on proper soil and water management, which creates the ideal conditions for crop development. Crop management starts with preparation of the land followed by planting or sowing of the planting material or seeds. Some crops, such as cowpea, pumpkin and maize, grow well if the seeds are sown directly in the garden. Others, such as tomato, onion and kale, are grown effectively by being raised first as seedlings in a nursery and transplanted later, while crops such as sweet potato and cassava are planted from vegetative material.

Before seeds are sown, or seedlings or cuttings planted, the land has to be prepared. The soil must be loosened to a depth of at least 20 cm. Tools suitable for this operation exist in most farming communities and are often made by the local blacksmith. Improved tools that reduce labour through the use of draught power or other forms of power besides human also exist but are often not accessible to the majority of African home garden managers.

The soil is formed into beds, ridges, mounds or other configurations suitable for the level of rainfall, soil type and its water-holding capacity, and the type of crop to be planted. When a plant is transferred from the seed bed to the permanent site, its root system should be disturbed as little as possible. Plants are dug up with a ball of earth attached or, if bare rooted, dipped in a slurry of mud. The best time for transplanting is late afternoon, to avoid direct sunlight which can damage the roots. Temporary shade can be made to reduce the sun's damage, especially in hot climates. After sowing or transplanting, water should be applied.

After sowing or transplanting, a plant needs care. Some plants need to be supported with stakes. Weeds must be removed, but roots and other plant parts must be protected from damage during weeding. The soil must be cultivated and its structure kept open, in order for air and water to penetrate. Unwanted plants should be thinned out, together with the weeds. Plant nutrients must be provided and pests controlled. Plant development must be carefully observed and progress monitored.

Depending on the part of the crop to be utilised, the point of crop maturity varies. Green leaves and shoots are harvested when still small and tender, while fruits and seeds need full development. Since crops do not always mature evenly, consecutive harvesting might be necessary. It is important to note that some plants do not develop new flowers or fruits if their mature fruits are not picked. This is because of the presence of growth inhibitors which are typical in some crops, such as pepper and avocado.

Post-harvest handling is extremely important, but is often the most neglected part of crop management. Poor post-harvest handling results in crop loss, which often can be reduced or avoided. When crops mature they should be consumed, processed or sold. After harvesting, a number of home garden crops need special treatment, such as curing. Curing reduces storage loss. The majority of crops needing curing are roots, tubers and bulbs (e.g. onion, sweet potato, Irish potato, cassava, taro and yam). Households may also cure pumpkin.

6

Cultivation of Site-specific Herbs

Profits and yields are essential for profitable production. An economic evaluation of the production of site-specific species within the sphere of a research project for the optimisation of seed production showed a satisfactory cover level for most of the species. Great care, preparedness for high risk and a learning process of several years are prerequisite. Nevertheless, the niche sector of the production of sitespecific seed varieties has its own rules.

The market is relatively limited. A lack of legal regulations still enables commercial mixtures of ecologically unsuitable species, which are significantly cheaper, to be used. Site-specific seed mixtures can only be sold through the intensive service and expertise of seed buyers.

Compared to cultivated species and varieties, all site-specific species have the common characteristic of an early slow development and little resistance to competition. All species need a perfectly prepared seedbed. Measures for plant protection must be undertaken as early as possible to avoid large deficiencies in yields. Biological planting is therefore out of the question for most species.

Many grasses show a high tendency to blight diseases. Only early control with suitable fungicides can prevent damage to the plant stand over a wide area. The regulations of the respective countries should be observed with regard to permission, use and precautionary measures when using pesticides.

Several of the sprays introduced have given good results in practice, but are not permitted for the relevant types of culture. Because weeds are one of the most significant problems in the production of site-specific seeds, the appropriate agents are at least named.

External qualities such as purity and germinating capacity are mostly within the producers' sphere of influence. A balanced species and timely supply of nutrition optimise the development and yield capacity of the species produced. Irrigation is indispensable in dry areas during the summer, and helpful at all sites in dry years.

The time and harvesting technique can decisively influence yield as well as the quality of the product. Drying facilities appropriately adapted for seed varieties are a prerequisite for the grasses, legumes and herbs described. Drying temperatures of 42°C should never be exceeded. Rapid transportation from the field to the drying facility is important for grasses above all because of the relatively high water content of the seeds and excessive independent heating of the threshed material.

Low humidity and low air temperatures are essential factors for long storage of seeds without damage.

Achillea Mllefolium s.l.

Euro-Siberian species with various ecotypes, very common; often found on soils that are nutrient-deficient to moderately nutrient-rich, on fields, along paths, even in semi-dry grasslands; on loam that is moist to moderately dry, medium-deep to deep, sandy, stony or pure.

Sensitive to long-lasting moisture or compactness of the topsoil as well as to deficient soil aeration. Extremely common in the Alps. All-purpose grass for restoration in lower altitudes up to the lower alpine region. Good tolerance of nutrients and cutting, resistance to trampling, not likely to be grazed by livestock.

Seeds

Achene 1.4-2.5 mm long, 0.8 mm wide, 0.3 mm thick; flat, thin like paper, oblongcuneate, cross-truncate at the apex. Hilum at the acute margin. Surface slightly vaulted, with fine longitudinal bands. Surface silver-grey to light greenish-grey, margin brighter, glossy. Thousand seed weight: 0.15-0.25 g.

There are no special requirements for soil and climate. Common yarrow prefers an intermediate to good supply of nutrients. The pH should range between 5.5 and 7.5. Wet, waterlogged and badly aerated soils should be avoided, as well as soils that are rich in peat and very light, dry soils.

Cultivation

Open sowing is possible, if sowing is carried out by the beginning of June at the latest (at mild locations by the middle of June). Crop irrigation is recommended. Sowing underneath low-density cover crops is also feasible.

Seed rate: 2-3 kg/ha.

Row spacing: 20-25 cm. Depending on the construction of the seeder and the method of weed control there should either be dense populations or sufficient row spacing (for mechanical and/or chemical weed control).

Fertilisation

For a satisfactory crop yield an intermediate amount of nutrients is required.

Phosphorus and potassium: on soils with an intermediate phosphorus and potassium content, basic fertilisation with manure (15-25 tonnes/ha solid or liquid manure) is sufficient in autumn. For mineral fertilisation 40-60 kg/ha P_2O_5 and 60-100 kg/ha K_2O are recommended.

Nitrogen: a sufficient supply of nitrogen in autumn guarantees satisfactory tillering.

The following amount is necessary for seed formation: 50-60 kg/ha of N-total to be added in autumn and early spring.

Maintenance and Weed Control

The species has intermediate juvenile development and produces little leaf mass. Thus, populations having low competition with weeds are only possible if mechanical weed control takes place (weeding, respectively brushing and hoeing between the rows), optimally combined with chemical weed control. As for the majority of herbs, focused chemical control of weed grasses poses no problems. To date there has been insufficient practical experience regarding the chemical control of dicotyledonous plants. But in summer and autumn of the seeding year and in spring of the harvesting year, it is possible to protect these plants by wiping wick and total herbicides (e.g. Glyphosate), thanks to the differences in growth heights between many weed grasses and common yarrow.

Harvest and Yields

Resistance to lodging: high.

Shattering tendency: low.

Ripeness: after flowering, the compound umbels change colour from dunnish, when ripe, to dark grey and dark brown. Until ripeness the seeds adhere fairly to the umbel quite well.

In order to check ripeness the inflorescence should be rubbed off on the palm of the hand: if sufficiently ripe, the silver-grey seeds will shed easily.

Ripening period: from the second to the third decade of July.

Red Flowering Ecotype

Harvesting technique: direct threshing is possible. The seeds are very small but have a good seed flow. During threshing the crop and sieves need to be checked frequently. The settings of the threshing concave should be narrow. Cleaning the seeds poses no particular problems.

Crop yields: yield forecasts are difficult, because of lack of experience concerning the cultivation of common yarrow. Yields range between 80 and 120 kg/ha. With proper experience, yields up to 100-150 kg/ha (and more) can be expected. Optimal maintenance leads to three harvesting years.

Anthoxanthum Odoratum

In the Alps this European species grows in hilly, montane and subalpine zones. Because of its early flowering period and ripeness, advancing to sites desiccating in summer is possible. This species grows on soil that is free of groundwater, and on almost any meadow land. As a result of its low growth, it prefers meadows.

Apart from rocks and rough debris the plant grows on almost any moist to semiarid soil with a low to intermediate nutrient supply and a pH between 3.2 and 6.5. This is an all-purpose grass in montane and subalpine sites, excellent for filling gaps thanks to the early ripeness of seeds and high seed production, its low biomass growth and low nutrient requirements. The grass is highly resistant to exhaust gases.

Seeds

Lemma of lower awned florets often apically glabrous with broad, membranous, white margins. Lemma of upper fertile florets at least on margins (often dorsally and the entire upper half) with short, stiff, patulous bristly hairs, often long cilia.

Glume oblong-elliptic, contracted, auburn, with long, shaggy, auburn hair; hairs pointing up; tips of glumes light yellow, translucent, membranous, glabrous, rounded. Awns of lower glumes 7-8 mm long, emerging from a little above the base, knee-high (same height as tip of glume), tightly contorted, black-brown, then thinner and lighter, tip of awn is whitish-yellow.

Inflorescence a panicle, 1.5-2 mm long, 0.6-0.8 mm wide and 0.3-0.5 mm thick, ovate-acuminate, can easily be squeezed out at glumes. Lemma smooth, auburn, shiny; palea finely scabrous, matt. Fruit 1-1.5 mm long, 0.5-0.7 mm wide, 0.3 mm thick, ovate-acuminate, tawny, matt.

Thousand seed weight: 0.5-0.6 g.

Light soil that is permeable and free of groundwater with a pH between 5.5 and 7.0 is suitable for seed production of sweet vernal grass. Propagation is not successful on compact, flat, waterlogged soil.

Cultivation

Sweet vernal grass needs a wellprepared, flat, fine seedbed. Seeds have a sufficient seed flow and are therefore suitable for all common seeding systems. Cultivation is possible by open sowing as well as under a cover crop. When using cover crops (e.g. summer cereals or linseed), it is important to have a light cover crop population because sweet vernal grass is not very tolerant to shade. Open sowing can be undertaken until the end of June or the beginning of July. Sufficient water supply is important.

Seed rate: 7-9 kg/ha are sufficient for good germination and optimal conditions for seed emergence.

Row spacing: 12-15 cm. Depending on the construction of the sower, broadcast sowing (e.g. by removing the sowing tubes) is also possible.

Fertilisation

Phosphorus and potassium: an intermediate supply of phosphorus and potassium is usually sufficient for propagation to develop successfully. Should additional fertilisation be required, 20 tonnes/ha of well-decomposed manure is recommended in autumn.

Nitrogen: main application of nitrogen should be carried out early in autumn in order to facilitate tillering and the development of spermatophors. Mineral fertilisation might prove necessary on poor soil and in dry weather in spring, in the event of bad nitrogen mobilisation. Sweet vernal grass develops quickly in spring. Later fertilisation at a disadvantageous stage of development (blossom, ripeness) facilitates the growth of leaves only and inhibits compact ripeness.

Maintenance and Weed Control

In the year of cultivation any maintenance has to be carried out with great care. Straw of the cover crop has to be removed and the stubbles cut. Effective weed control in the first year is the basis for a population that is free of weeds and has a high crop yield.

As for most grasses the following guidelines apply. Before sowing, only contact herbicides should be used, up to three days before the beginning of sowing. Hormone-type and broad-spectrum herbicides should generally not be used until the species has reached the three leaf stage. Note that the herbicides listed in Table 3 are not registered for use in seed cultivation in some countries. Especially when cultivated in lowlands or in humid weather, sweet vernal grass is very sensitive to rust and leaf diseases.

Application of fungicides: (usage and dose as in grain cultivation) may, therefore, already be necessary in late summer of the first year or specifically after the first harvest. Note the instructions for the use of herbicides given in the introduction.

Cutting in autumn is only necessary for high growth. Sweet vernal grass is resistant to snow mould and low temperatures in winter. Especially with high rust infestation, cutting in late summer or in spring of the first harvest year might be useful or even necessary. Because of early ripeness, competition with weeds may be very high in the summer of the first harvest year. Selective weed control with wick and total herbicides (e.g. Glyphosate) is possible because of different growth heights.

Harvest and Yields

Resistance to lodging: high.

Shattering tendency: very high.

Ripeness: seeds assume a typical brown colour.

Panicles change colour to golden yellow, similar to straw. Seeds ripen one after the other from the apex down to the base. Defining the optimal time of ripeness is extremely difficult.

Ripening period: very early, in warm regions in the middle of May, in cooler regions in the last week of May to the beginning of June. There are important differences between ecotypes. A more compact ripeness can be achieved by appropriate selection.

Harvesting technique: because of the plant's high shattering tendency, direct threshing is recommended. The rotational speed of the threshing drum should be low to avoid dehusking of seeds. Seeds as well as chaff in the sieve should be checked several times while threshing.

Seed Propagation Stand

Settings of the threshing concave: generally close; the drier the seeds, the wider the setting in order to prevent seed damage.

Crop yields: crop yields can vary greatly, between 50 and 250 kg/ha. As for all site-specific species, specialised knowledge and several years of experience are necessary to obtain optimal, stable crop yields.

Only two harvesting years are possible because of the intensive maintenance required. Site-specific grasses and herbs—Seed production and use for restoration of mountain 13

Anthyllis Vulneraria

This central and south European species occurs in the foothills of the Alps, the montane and—as with subspecies alpestris—the subalpine zone. In the valleys it grows mainly on semiarid grasslands; on mountain slopes on calcareous, poor grasslands. It prefers skeletal brown soils, pararendzina, rendzina, stony brown and red loam and initial and bare soils. Found in the Alps up to approximately 1 800 m.

Seeds

This species is suitable for restoration between the foothills and subalpine zones, especially on semiarid and stony sites, combined with low-growing grasses. It protects the soil against surface and depth erosion by means of its leaves, which grow close to the ground, and its extensive root system. Tends not to grow in mass and does not require many nutrients.

Seeds 2-3 mm long, 1-2 mm wide and 0.8-1.2 mm thick; broad-ovoid, sometimes with a slight lateral indentation. Radicle clinging, not sticking out. Hilum centrally ventral, round, whitish, with dark border. Surface colour significant: yellow up to two-third, remaining part green; glossy, smooth. Pod one-seeded, inflated oblique-ovoid, pilose, dark to black-grey.

Thousand seed weight: 2.3-2.7 g.

Common kidney vetch is extremely undemanding and adaptable, while tolerating fertilisation intermediately; even suited to poor habitats with shallow, nutrient-deficient soils. Sites that are too damp and clay and peat soils are not suitable. The pH should not be above 6, otherwise there may be a deficient root development. Kidney vetch prefers arid weather, having a low transpiration coefficient. It has high nitrogen accumulation capabilities, is self-tolerant and a good previous crop for cereals, sweetcorn and rape. As for all legumes, there should be a long crop rotation in order to prevent an accumulation of fungal pathogens.

Cultivation

Common kidney vetch needs a fine, crumbly and compact seedbed. Open sowing is possible throughout the year. In order to guarantee a satisfactory crop yield the following year, seeds should be sown up to July at the latest. According to early literature on seed cultivation of forages, autumn sowing underneath rye is possible in September. Common kidney vetch is suitable as a nurse crop for summer barley or durum. It is not troublesome at sowing.

Seed rate: 8-10 kg/ha.

Row spacing: 12-25 cm, depending on the construction of the seeder.

Fertilisation

Phosphorus and potassium: common kidney vetch requires very few nutrients. In autumn before cultivation, basic fertilisation with manure (15-20 tonnes/ha) is considered sufficient. Using mineral fertiliser, an application of 60-80 kg/ha P_2O_5 and 100-140 kg/ha K_2O is recommended, depending on soil content.

Nitrogen: kidney vetch meets its nitrogen requirements by means of nitrogen-fixing bacteria. As for all legumes, fertilisation with 20-30 kg N is recommended for open sowing in order to stimulate juvenile development.

Maintenance and Weed Control

In early autumn before the crop year, cutting is recommended if populations are densely grown. Having low competition with weeds, mechanical weed control should be combined with the application of herbicides. Kidney vetch is not sensitive to the use of curry combs. As for all legumes, specific weed control with herbicides is not always possible. In this context, experiences in clover seed production should be mentioned. However, compared to most clover species, kidney vetch does not tolerate herbicides containing butyric acid (MCPB).

Flowering Plants in the Second Vegetation Period

Pests and diseases: common kidney vetch is usally healthy, although several fungus diseases (mildew, focal spot disease, rust and stem fungal) are possible. In the north of Italy, root rot caused by fungal pathogens appears from time to time, which can lead to heavy losses.

Most kidney vetch plants die off after seed ripeness. Plant survival depends on origin, ecotype and sowing time. A second harvest the following year can be profitable if the reproductive populations together with the juvenile plants—grown through seed shedding—remain sufficiently dense. In this case, densely grown populations have to be cut in autumn in order to minimise losses over winter.

Harvest and Yields

Resistance to lodging: intermediate to low, depending on origin.

Shattering tendency: extremely high.

Ripeness: at time of ripeness florets change colour to silver-grey and pods become dark to black-brown. Seeds become firm (nail test) and change colour to a typical green-yellow. Because of the plant's unequal ripening and high shattering tendency, harvesting is only recommended when at least 60 percent of the flowers are ripe. Often this state of ripeness is reached when the third axis of flowers at the stem is ripe for threshing.

Ripening period: starting in the middle of July and lasting up to the middle of August, depending on origin.

Harvesting technique: experiments show conspicuously higher crop yields with direct threshing.

Swath threshing is not recommended because of high losses through seed shedding. Threshing has to be carried out with great care.

Crop yields: yield forecasts amount to between 200 and 400 kg/ha. Depending on ecotype and weed infestation a second harvest is possible.

Depending on the thresher, harvested seeds can still remain covered with 60 to 90 percent of pods. As is common practice in seed production, seeds should be dried gently immediately after harvesting. Hulling the seeds should be carried out with special machines (clover rubber).

ANTHYLLIS VULNERARIA SSP. ALPESTRIS

Alpine kidney vetch

This plant of the central and southeast European mountains occurs on similar sites to Anthyllis vulneraria. It is very common in the Alps, but only in the subalpine and alpine zones. The species prefers calcareous soils, but is also found on lime-deficient ground. On calcareous, poor grasslands it sometimes represents the stand structure. The species is also frequently found on talus sites; it grows in the Alps at altitudes of up to 3 000 m. On calcareous ground the plant is particularly suitable for restoration of slopes and rocky, planed soils of the subalpine and alpine zones.

Seeds

Seeds 2.4-3.2 mm long, 1.6-2 mm wide and 0.8-1.2 mm thick; broad-ovoid, sometimes with a slight lateral indentation.

Radicle clinging, not sticking out. Hilum centrally ventral, round, whitish, with dark border. Surface colour significant: yellow up to two-thirds, remaining part green; glossy, smooth. Pod one-seeded, inflated oblique-ovoid, pilose, dark to blackgrey.

Thousand seed weight: 3.2-3.5 g.

Alpine kidney vetch is extremely undemanding and adaptable, while tolerating fertilisation intermediately. Even poor habitats with shallow, nutrient-deficient soils are well suited. Sites that are too damp and clay and peat soils are not suitable. The pH should not be below 6, otherwise a deficient root development may form. Alpine kidney vetch prefers arid weather having a low transpiration coefficient. It is self-tolerant and a good previous crop for cereals, sweetcorn and rape. As for all legumes, there should be a long crop rotation in order to prevent an accumulation of fungal pathogens.

Cultivation

Alpine kidney vetch should be threshed close to the ground. A flat, fine, crumbly seedbed without stones is preferred. Open sowing is possible throughout the year. In order to guarantee a satisfactory crop yield the following year, seeds should be sown up to July at the latest. Alpine kidney vetch is extremely suitable as a nurse crop for summer barley or durum. It does not cause any problems during sowing.

Seed rate: 8-10 kg/ha.

Row spacing: 12-25 cm, depending on the construction of the seeder.

Fertilisation

Phosphorus and potassium: alpine kidney vetch requires very few nutrients.

In autumn before cultivation, basic fertilisation with manure (15-20 tonnes/ha) is considered sufficient. Using mineral fertiliser, an application of 60-80 kg/ha P_2O_5 and 100-140 kg/ha K_2O is recommended, depending on soil content.

Nitrogen: kidney vetch meets its nitrogen requirements by means of nitrogen-fixing bacteria. As for all legumes, fertilisation with 20-30 kg/N is recommended for open sowing in order to stimulate juvenile development.

Maintenance and Weed Control

Compared to common kidney vetch, alpine kidney vetch has a slower juvenile development and produces conspicuously less leaf mass. Thus, populations with low competition with weeds are only possible with mechanical weed control, optimally combined with chemical controls. The plant is not sensitive to the use of curry combs. As for all legumes, specific weed control with herbicides is not always possible. In this context, experiences in clover seed production should be mentioned. However, compared to most clover species alpine kidney vetch does not tolerate butyric acid (MCPB).

Pests and diseases: *Anthyllis vulneraria ssp.* alpestris is usually healthy, although several fungal diseases (mildew, focal spot disease, rust and stem fungal) are possible. In the north of Italy, root rot caused by fungal pathogens appears from time to time, which can lead to heavy losses.

Most alpine kidney vetch plants die off after seed ripeness. Plant survival depends on origin ecotype and sowing time. A second harvest the following year can be profitable if the reproductive populations together with the juvenile plants—grown through seed shedding—remain sufficiently dense. In this case, densely grown populations have to be cut after the first harvest in autumn in order to minimise losses over winter.

Harvest and Yields

Resistance to lodging: low.

Shattering tendency: extremely high.

Ripeness: at time of ripeness florets change colour to silver-grey and pods become dark to black-brown. Seeds become firm (nail test) and change colour to a typical green-yellow. Because of the plant's unequal ripening and high shattering tendency, harvesting is only recommended when at least 60 percent of the flowers are ripe. Often this state of ripeness is reached when the third axis of flowers at the stem is ripe for threshing.

Ripening period: about one week earlier than common kidney vetch, mostly in the first decade of July.

Harvesting technique: because of limited leaf mass the stems lie close to the ground and therefore the board has to be kept very low. Swath threshing is not recommended because of high losses through seed shedding. Threshing has to be carried out with great care.

Crop yields: yield forecasts amount to between 100 and 300 kg/ha. Depending on ecotype and weed infestation a second harvest is possible.

Depending on the thresher, harvested seeds can still remain covered with 60 to 90 percent of the pods. As is common practice in seed production, seeds should be dried gently immediately after harvesting. Hulling the seeds should be carried out with special machines (clover rubber).

Avenella Flexuosa

Wavy hair grass

This plant occurs worldwide and in the Alps it grows in the foothills and the montane, alpine and subalpine zones. In the foothills and montane zones it needs shade and therefore grows mainly in forests. In the high montane and subalpine zones it tolerates or requires full exposure. There it occurs in glades, meadows and dwarf shrub heaths. The species is sensitive to airdeficiency in soils. Its natural habitats in Europe are generally nutrient-deficient. The type of soil is usually fustyhumus brown soil, sandy to loamy brown soil, podsol brown soil, podsols or desiccated peat. The plants grows on soils with a pH between 4 and 7, although a pH of 3.3-3.5 is sometimes tolerated. In the Alps it grows up to 2 750 m above sea level.

Important grass for soil with fine-earth deficiency and humus, acidic, arid sites above the timber line as well as halfshaded forest aisles and slopes in lower altitudes. It prefers nutrient-deficient soil and is sensitive to foraging.

Seeds

Fruit of the husk 25-35 mm long, 0.6-0.9 mm wide, 0.5-0.8 mm thick, ovatelanceolate. Awn meandering, auburn, slightly longer than lemma, growing from above the base. Lemma membranous, silvery translucent, longer than fruit, lemma transcending margins of palea. Palea flat, in narrow hollow, flimsy. Base with few white hairs. Stipellus short with few but long hairs. Surface tawny to brownis matt. Fruit 2-3 mm long, 0.5-0.7 mm wide, 0.5-0.6 mm thick, oblong, brown, matt.

Thousand seed weight: 0.5-0.7 g.

This species prefers arid or moist sites up to 1 000 m above sea level. Compact and air-deficient soils should be avoided. Soils should be water permeable and rich in humus. In the literature there are reports of growth disorders and chlorosis when the pH of the soil is above 6. However, in cultivation experiments on soils with pH 7 this was not the case. On arid sites irrigation is recommended in order to facilitate a regular water supply. Wavy hair grass is, however, sensitive to waterlogged soil and soil that is blocked by alluvial deposits, especially in its juvenile development.

Cultivation

Wavy hair grass should preferably be sown in spring with broadcast sowing no deeper than 0.5 cm. Cultivation below winter barley is not recommended because the species has a slow juvenile development. Satisfactory development of single plants until autumn will therefore not be possible. Seed flow is sufficient for common seeding systems. A light cover crop population is important because of slow juvenile development. Summer barley and linseed have proved worthwhile in practice.

Seed rate: 10-12 kg/ha.

Row spacing: for this species, row spacing should be between 12 and 15 cm. Broadcast seeding is possible.

Fertilisation

Phosphorus and potassium: wavy hair grass prefers nutrient-deficient soils. For an intermediate supply of phosphorus and potassium additional fertilisation is not necessary. In autumn small amounts of manure can be applied without risk.

Nitrogen: this too should only be applied in small amounts. A moderate application of nitrogen in autumn and/or in early spring (no more than 20-40 kg/N each, depending on climate) facilitates a good yield. On soils with a pH higher than 6, chlorosis may occur; the application of ferruginous fertilisers (not as sulphate) will then be necessary.

Maintenance and Weed Control

Because of slow juvenile development, intensive maintenance is necessary and weed control has to be carried out in the weeks after harvesting the cover crop. Wavy hair grass does not grow very high in the summer of cultivation (0.5-1.5 cm) and easily risks being covered by crop residues. Therefore straw has to be removed immediately. Chaffing and spreading crop residues lead to plant losses on large areas.

Single Plant in a Propagation Stand

As for most grasses the following guidelines apply. Before sowing, only contact herbicides should be used, up to three days before the beginning of sowing. Hormone-type and broadspectrum herbicides should generally not be used until the species has reached the three leaf stage. It should be noted that the herbicides listed in Table 3 are not registered for use in seed cultivation in some countries.

The low growth height of this species requires considerable effort in maintenance and weed control. Selective grass weed control is difficult. Because of different growth heights of wavy hair grass and other grasses, selective control with wick and total herbicides (e.g. Glyphosate) is possible within certain periods of the growing season.

Harvest and Yields

Resistance to lodging: high.

Shattering tendency: low to intermediate.

Ripeness: the axes of the panicles change colour to light tawny. Seeds adhere well to the panicle after ripeness and change colour from greenish-brown to brown. To check ripeness, a bunch of panicles should be dutched firmly: the brown seeds will fall off easily if ripe.

Ripening period: last decade of June or first decade of July. Because of the low shattering tendency it is easy to fix the time of threshing.

Harvesting techniques: wavy hair grass can easily be threshed directly. The rotational speed of the threshing drum should be set low to avoid mechanical damage or dehusking of seeds. Seed flow is sufficient. Seeds as well as the chaff in the sieve should be checked several times while threshing.

Settings of the threshing concave: generally close, the drier the seeds, the wider the setting in order to prevent seed damage. Cleaning of seeds usually presents no problems.

Crop yields: crop yields can vary greatly (12-150 kg) where there is little experience in propagation. Where experience is considerable 100-150 kg can be expected. Appropriate prices for seeds are a precondition for economically profitable production.

For optimally maintained populations, three harvest years are possible.

Bellardiochloa Variegata

Violet meadow grass

In arid grasslands on sunny, warm slopes, pioneer plant in crevices, on ledges and cornices. On dry soil that is free of lime, neutral to slightly acidic, nutrient-deficient, stony to rocky soil. Indicates aridity, poorness and light (heliophyte). Preferably in the subalpine, seldom in the alpine zone. All-purpose grass for all silicious sites or sites deficient in lime, as far as the lower alpine zone. Tolerant of nutrient but not dependent on it, tolerant of cutting to a certain extent, not tolerant to grazing.

Seeds

Glumes almost similar, three-veined, 3-4.5 mm long, lanceolate, acuminate, keeled, midvein scabrous. Lemma five-veined, 3.6-4 mm long, slightly keeled, apically acuminate, at least lower ones expiring in 0.3-1 mm long awns. On lower half of keel and marginal veins short hair, florets without woolly hair on base. Palea two-veined, almost as long as lemma, oblong-lanceolate, apically slightly notched; in-between keels short hair, keels with very short, acute, bristly hair. Anther 1-1.8 mm long, smooth and glabrous, outline elliptic.

Thousand seed weight: 0.3-0.5 g.

Violet meadow grass has no special soil requirements. Moist or waterlogged soil, ground with high weed infestation, peat and light, dry soil should be avoided. Competition with weeds is low. Locations with a high presence of annual meadow grass (Poa annua) rough meadow grass (Poa trivialis) and, above all, smooth meadow grass (Poa pratensis) should be avoided at all costs. Propagation is possible with a pH between 5.0 and 7.0.

Cultivation

Violet meadow grass needs a fine, crumbly seedbed. Open sowing is possible but risky because of the danger of high weed infestation. For the sufficient development of single plants until autumn, open sowing must be carried out until the middle of June. For open sowing in summer, irrigation must be possible.

Seeding underneath summer barley/durum wheat is common. A thin cover crop population is important. If stands are too dense, the development of single plants is suppressed, which leads to bad tillering in autumn. Furthermore, increased weed competition and a severely decreased crop yield may occur. Good experience has been recorded with linseed as a cover crop. Because of the plant's slow juvenile development, seeding should be done immediately after seeding the cover crop. Seeds have a good seed flow and are suitable for all common seeding systems. It is important that seeds are applied flat and evenly in a maximum depth of 0.5 cm, and rolled afterwards.

Seed rate: 8-10 kg/ha with an excellent seeding technique, otherwise higher amounts are required to obtain regular stands.

Row spacing: 15-20 cm, depending on the seeder.

Fertilisation

Violet meadow grass is fairly undemanding. For satisfactory seed development a good supply of nutrients is necessary.

Phosphorus and potassium: on soil with an intermediate supply of phosphorus and potassium early fertilisation with manure in autumn is recommended. For soils with sufficient nutrient content 50-70 kg/ha P_2O_5 and 80-120 kg/ha K_2O are necessary.

Nitrogen: after harvesting the cover crop and cutting the stubbles, an application of 30 kg/ha/N-total is recommended, to achieve optimal development until autumn. Nitrogen necessary for seed development should be applied in autumn or in early spring, for an amount of approximately 70 kg/ha/N-total. Late fertilisation in spring mainly stimulates the development of leaves.

Maintenance and Weed Control

This species grows quite high and forms rather dense, enduring stands after having outgrown the juvenile phase. Herbicides are well tolerated. Because of low competition, weed control has to start as early as possible to obtain a satisfactory development.

A special problem is weed infestation by other meadow grasses. Selective weed control is only possible for *Poa annua*. Smooth meadow grass in particular can overgrow the entire population within two years and cause problems with yields and quality as a result. The only effective weed control is to use a wick for the control of single plants with total herbicides (Glyphosate).

Sensitivity to rust: quite low. Rust control will only be necessary for a few years.

There should be no cutting between the harvest and autumn in order to minimise the development of weeds. Violet meadow grass is not used as forage.

Harvest and Yields

Resistance to lodging: intermediate.

Shattering tendency: intermediate.

Ripeness: branches of panicle and upper parts of culms change colour to golden yellow. Seeds shatter easily when touched. At the time of ripening the stands should be checked up to twice a day because ripeness is not very compact.

Ripening period: usually shortly after alpine meadow grass in the last decade of June.

Threshing is always done from the root. The cutter bar of the harvester can be set fairly high. Settings for threshing are usually unproblematic. The rotational speed of the threshing drum can vary between 800 and 1 000 rpm with close settings of the threshing concave. The fan should be set very low-usually the wind produced by the threshing drum is sufficient.

Crop yields: crop yields are usually higher in the second harvest year. In existing seed production the average crop yield has been about 200 kg/ha. The yield potential is approximately 400 kg/ha. Maintenance and weed control are crucial for a satisfactory yield level.

Briza Media

Perennial quaking grass, quaking grass

In the Alps this European species occurs in hilly, montane and subalpine zones. At lower locations mainly in thin populations, at higher altitudes in densely grown turfs also. In nutrient-deficient meadows and pastures, along road embankments, on rocky slopes, and in low-density stands. On clay soil, loam and sandy loam that is moderately dry or moist up to wet, alkaline, often nutrient-deficient, moderately acid soil, rich in humus or peat, with a pH of 3.8-7.2. Indicating nutrient-deficiency and light (heliophyte), shallow-root plant on humus. Good but low-productive forage. In meadows and pastures without fertilisation the plant is stimulated by maintenance, but disappears with intensified fertilisation and grazing.

All-purpose grass for restoration in montane to alpine regions, less qualified for highly acid and moist soils; cespitose, preferably for poor habitats, intermediate forage quality.

Seeds

Glumes similar, three to five-veined, 2.5-3.8 mm long, broadly ovate to orbiculate as seen from dorsal view, upper margin rounded, vaulted, membranous, margins hyaline, glabrous, midvein sparsely scabrous.

Lemma seven to nine-veined, 3.6-4.2 mm long, similar shape to glumes, densely arranged, membranous, green, often coated purple, margins hyaline and whitish. Palea two-veined, 3.5-4 mm long, flat, ovate, apex slightly nicked, keels prominent and narrow-winged. Anther 2-2.5 mm long. Fruit about 2 mm long, broadly elliptic in contour, light brown.

Thousand seed weight: 0.3-0.5 g.

Warm, moist to moderately dry soils with a pH of 5.0-7.5 are suitable for seed gain cultivation. Compacted waterlogged soils should be avoided.

Cultivation

Like most grasses, perennial quaking grass needs a carefully prepared, fine, crumbly seedbed. Seeds have sufficient seed flow; however, they can be easily blown away by the wind.

This species is relatively fast-growing compared to most of the species described. It can be sown underneath spring corn, blank seed is also possible. Attention should be paid to low-density cover crop populations. Open sowing should be carried out by the end of June at the latest, in order to allow sufficient tillering in autumn. In this case, it is recommended that the crop is irrigated during summer droughts.

Crop yields: Under average conditions, optimal maintenance allows for yields between 120 and 300 kg/ha.

Seed rate: 10-12 kg/ha is common.

Generally speaking, a decrease in the seed rate can help to produce welldeveloped and well-tillered single plants. However, this is only recommended where there is expert knowledge and best population tending.

Row spacing: 15-20 cm, depending on the construction of the seeder. Broadcast sowing is also possible in combination with a curry comb.

Fertilisation

Phosphorus and potassium: although quaking grass requires few nutrients, moderate fertilisation (40-60 kg/ha P_2O_5 and 80-100 kg/ha K_2O) works well. Fertilisation with liquid or solid manure (in late summer and early autumn) stimulates the growth the single plants until their rest period. Application of liquid manure in spring should be avoided because of nitrogen mobilisation during an unfavourable vegetation stage.

Nitrogen: this should be applied in early spring, only with mineral fertiliser, in amounts of 50-70 kg/ha depending on soil and climate.

Maintenance and Weed Control

The straw of the cover crop should be removed immediately and then cut. As for most grasses the following rules apply. Before sowing, only contact herbicides should be used, preferably up to three days before the beginning of sowing. Hormone-type and broadspectrum herbicides should generally not be used until the species has reached the three leaf stage. It should be noted that the herbicides are not registered for use in seed cultivation in some countries.

Having tended the species well after the cover crop harvest, densely grown populations can develop until autumn. Where heavy growth of weed grasses is evident, the specific application of Ethofumesate can control Poa annua in particular. However, an expert should be responsible for evaluation as well as composition. National guidelines on the use of herbicides should always be followed.

Rust diseases: perennial quaking grass is extremely sensitive. Especially after the first seed harvest in summer, the population may be infected within a couple of days and collapse as a result. Continuous monitoring and—weather permitting—prophylactic spraying with a suitable fungicide are advisable.

If the crop is already infected by rust, it should be cut and removed immediately, and combined with spraying about one week after the cutting in order to save the remaining population. Crop yield losses need to be reckoned with the following year.

Harvest and Yields

Resistance to lodging: intermediate.

Shattering tendency: intermediate.

Ripeness: culms and flower spikes change colour from reddish light brown to tawny or light brown. Until complete ripeness is reached, there is a low shattering tendency, which then becomes high. As for most grasses appropriate to the habitat, it is very important to keep an eye the ripening in order to optimise crop yield and quality.

Flowering Propagation Stand

Ripening period: between the end of June and the beginning of July, depending on climate and ecotype.

Harvesting technique: direct threshing is possible and easy. Because of a high accumulation of biomass, the straw from the harvest should at least be chaffed or carried away. More than for other species, seed producers should set the threshing drum at a low rotational speed, with a comparatively wide gap between drum and threshing concave. Large sieves are needed. In order to avoid high losses at threshing, the air supply should be reduced, if necessary to zero. Seeds should be cleaned with great care. Cleaning with air and adequate sieve systems is normally sufficient.

Crop yields: Under average conditions, optimal maintenance ensures yields between 120 and 300 kg/ha.

Bromus Erectus

Upright brome, erect brome

This European species occurs in the Alps, in the foothills and the montane zones. It requires a temperate, highly maritime climate. Upright brome grows mainiy on dry, lime-rich soils. But in the Mediterranean region it prefers shady sites and moister soils. It grows also on rendzina, pararendzina, calcareous brown soils and alkaline ranker with a pH of 6.3-7.4.

General-purpose grass for restoration of drier, alkaline, nutrient-deficient sites at intermediate altitudes. Partial tolerance of cutting, sensitive to grazing and fertilisation.

Seeds (Blaschka, A.)

Fruit 9-11 mm long excluding awn, 1.5 mm wide, and 1-1.2 mm thick.

Lemma narrowly oblong, apically and basally acuminate, dirty straw coloured, dorsally with short or long hairs, partly glabrous.

Awn 4-6 mm long, erect, originating below the apex of the lemma. Palea irregularly spaciously ciliated at both keels, flatly grooved, a little shorter than the lemma, hyaline, colourless. In comparison to Bromus inermis the keels have longer teeth, with shorter ones lying in between. Rachilla 1.5-3 mm long, straight, round, becoming thicker towards the apex, often with short hairs. Fruit adhering, separable from the husks only with effort, 7-9 mm long, 1.5-2.5 mm wide and 0.5-0.7 mm thick.

Thousand seed weight: 3.2-5.5 g.

Like Bromus inermis, upright brome is extremely undemanding. Intermediate to light soils that are rich in humus and drier sites are suitable. It is important to select sites with little weed competition because of limited possibilities for the use of specific herbicides.

Cultivation

This species needs a well-prepared seedbed. Open sowing is recommended (this is possible until the end of June if irrigation is available). Upright brome is moderately tolerant if grown underneath a cover crop. Therefore populations of spring corn should be low density; linseed oil flax is recommended.

Seed depth: 1.5-2 cm maximum.

Seed rate: 12-15 kg/ha.

Row spacing: 25-30 cm. Broadcast sowing is also appropriate.

Fertilisation

Upright brome requires an intermediate amount of nutrients.

Phosphorus and potassium: on soils with an intermediate phosphorus and potassium content, basic fertilisation with manure (liquid or solid, 15-25 tonnes/ ha) is sufficient in autumn. Depending on soil content, fertilisation with 60-80 kg/ha P_2O_5 and 80-120 kg/ha K_2O is considered sufficient, depending on the yield.

Nitrogen: after harvesting the cover crop, or at sowing, fertilisation with 30-50 kg/ha of N-total is recommended so as to obtain best development until autumn. For harvest growth, 70-80 kg/ha/y of Ntotal is necessary; to be added in autumn and early spring. If spring fertilisation is carried out too late, the main result is development of the leaf mass.

Maintenance and Weed Control

There are special problems connected to the control of undesirable brome species. Bromus tectorum or B. sterilis accumulate especially on drier sites and it is complicated to control them.

Dense populations grow extensively until autumn of the seeding year. The timely and consistent use of herbicides is vital. Specific control of weed grasses can only be carried out by controlling single plants, or in favourable years by the temporary use of wiping wicks.

Harvest and Yields

Resistance to lodging: high.

Shattering tendency: low.

Ripeness: the spike changes colour to brown and appears to dry up. The ripeness of the seeds is determined by the nail test. Culms and leaves are still green at this point.

Ripening period: from the middle until the end of July.

Harvesting technique: direct threshing is recommended. Seeds have sufficient seed flow and can be cleaned easily.

Crop yields: yields depend largely on the cultivator's experience. The range is between 200 and 500 kg/ha. Optimal maintenance ensures two to three harvesting years.

Cynosurus Cristatus

Crested dog's tail, dog's tail grass

In the Alps this perennial Eurasian species occurs from the foothills to the subalpine zone. In the lower valleys of the Alps that are warm in summer, the plant decreases significantly. At these locations dog's tail grass prefers sites with good water support and less sun. In the montane zone with its more balanced climate, the plant grows more often in meadows and pastures. There it also appears in natural and artificial turf sites, in country lanes and in low-density stands.

On clay soil and loam that is moist, rich in nutrients, alkaline, moderately acid, mostly compacted, indicating light (heliophyte). It likes moist, moderately rich in nutrients and humus, loamy-clayey soils with a pH of 5-7. It is a good forage grass that protects the soil against erosion thanks to its good resistance to intensive grazing (sheep pastures). Good nutrient tolerance (with the exception of high amounts of nitrogen), suitable for restorations up to the subalpine zone as well as in lower regions. Dog's tail grass is an intermediate forage that is also used as a component in grassland mixtures.

Seeds

Glumes similar, one-veined, 3-4.5 mm long, narrowly lanceolate as seen from a lateral view, acuminate, hyaline, glabrous, with a conspicuous scabrous keel. Lemma five-veined, 3-4 mm long, lanceolate, apex narrowly rounded and mostly with a thin awn, up to 1 mm long, membranous, margins hyaline; glabrous, scabrous at upper part. Palea about as long as the lemma, lanceolate, apex nicked, scabrous on the keels.

Thousand seed weight: 0.5-0.7 g.

Seed production is possible on all kinds of productive soils. Soils that are rich in nutrients and have good water management are preferred. However, heavily waterlogged or very light sandy soils are not suitable.

Cultivation

Dog's tail grass is fairly tolerant to shade and therefore can be sown underneath winter wheat as well as spring corn. Open sowing should be carried out by the middle of June at the latest. Effort concerning weed control is needed.

Seed rate: 7-9 kg/ha are sufficient if growing conditions and germination capacity are satisfactory. If sown underneath a cover crop in autumn or blank in spring, the seed rate should be slightly increased.

Row spacing: 15-20 cm. Broadcast sowing (e.g. by removing sowing tubes) is also possible in combination with a curry comb.

Fertilisation

Phosphorus and potassium: although dog's tail grass requires few nutrients, a moderate fertilisation (40-60 kg/ha P_2O_5 and 80-100 kg/ha K_2O) has shown good results. Timely fertilisation with liquid or solid manure (late summer, early autumn) stimulates the growth of the single plants until their resting period. Application of liquid manure in spring should be avoided because of undesirable nitrogen mobilisation during an unfavourable vegetation stage.

Nitrogen: total requirement per harvesting growth is 70-80 kg/ha N. The plant should be fertilised in spring, only with mineral fertiliser, in amounts between 40 to 60 kg/ha depending on climate and weather.

Maintenance and Weed Control

Dog's tail grass has a slow initial development. Even if there is more fertilisation, only a little leaf mass will be produced. Thus, as a rule, additional cutting is necessary only in autumn. Particularly at lower locations, dog's tail grass is sensitive to rust diseases. In this case it is essential to use fungicides or carry out early cutting.

Before sowing, only contact herbicides should be used, up to three days before the beginning of sowing. Hormone-type and broad-spectrum herbicides should generally not be used until the species reaches the three leaf stage. Until autumn of the year of sowing and after the seed harvest, specific weed control can be successfully carried out with wiping wick and total herbicides (Glyphosate).

Harvest and Yields

Resistance to lodging: high.

Shattering tendency: intermediate.

Ripeness: the culm changes colour to tawny and the pseudo-spike goes from light brown to a coffee colour. When unripe the seeds are stained bright greenish-yellow, becoming more and more tawny during ripening. Even at the beginning, the seeds can be rubbed off quite easily, although they are relatively firm.

There is a risk of flailing too early: in order to estimate the ripeness of the seeds, a nail test is advised. The seeds should at least be at the yellow ripeness stage.

Ripening period: does not flower too late; ripeness depends to a great extent on the weather. On average, it can be expected within the last decade of July.

Harvesting technique: direct threshing is recommended, which is very easy because of the high resistance to lodging. The seeds have an excellent seed flow and can be cleaned easily.

Crop yields: optimal maintenance allows yields between 200 and 600 kg/ha. Up to three harvesting years are possible.

Deschampsia Cespitosa

Tufted hair grass

This species occurs worldwide in temperate humid regions. In the Alps it can be found in the foothills and the montane, subalpine and alpine zones. Because of its deep roots, the species tolerates occasional drought in the topsoil. Its pronounced ability to crowd out other species results in generous growth. Deschampsia cespitosa var. alpina prefers depressions, slopes with soakage and sites around the leakage of spring water. The species grows on soils with a pH of 4.3-8.1 and grows at altitudes of up to 2 750 m in the Alps.

With regard to altitude, there is considerable ecological amplitude. This is the most important grass in the restoration of damp to alternating moist and dry locations. After restoration, spontaneous dispersal to sites appropriate to its habitat is often noted. Good stabilisation of the soil and tolerance of nutrients; at higher locations it is grazed by livestock, but its forage value is low. On pastures it is controlled because it is a weed and also takes up space; different control methods are used.

Seeds

Fruit 1.5-2.5 mm long, 0.6-0.8 mm wide, 0.5-0.6 mm thick; ovate-lanceolate. Awn meandering, auburn, slightly longer than the lemma, arising almost basally. Lemma hyaline, translucent, about twice as long as the fruit, dentate at the top; margins bent. Dense white hair tuft at base.

Rachilla: with white hair. Fruit 1.5 mm long, narrowly acuminate, brown.

Thousand seed weight: 0.2-0.4 g.

Seed cultivation is possible on any soil that is well supplied with water and has a pH range from 5.0 to 7.5. Tufted hair grass also tolerates compact and waterlogged soils, where farming is still possible. The species is hardy and even tolerates strong frost. Provided water supply is sufficient, fertility is satisfactory.

Cultivation

Tufted hair grass needs a well-prepared, fine, crumbly seedbed. It is sensitive to seed depths exceeding 0.5 cm. Seeds have sufficient seed flow and can be sown with an established seeder. It is advisable to sow underneath spring corn or blank until the beginning of June, in order to guarantee the sufficient development and tillering the single plants. When sown on bare soil, attention should be paid to irrigation.

Seed rate: 6-8 kg/ha is sufficient for dense populations.

Row spacing: 15-20 cm, depending on the construction of the seeder. Broadcast sowing is also possible.

Fertilisation

This species is robust and undemanding. Nevertheless, a

good supply of nutrients is recommended in order to optimise the crop yield.

Phosphorus and potassium: On soils with an intermediate phosphorus and potassium content, basic fertilisation with manure (liquid or solid 15-25 tonnes/ha) is sufficient in autumn. For mineral fertilisation, 40-60 kg/ha P_2O_5 and 60-100 kg/ha K_2O are recommended.

Nitrogen: a sufficient supply of nitrogen in autumn guarantees a satisfactory tillering. It should be borne in mind that an excessive supply of nitrogen in late spring may cause decreased formation of the spermatophores.

However, a certain amount of nitrogen is necessary for the formation of seeds: 50-70 kg/ha of N-total should be added in both autumn and early spring.

Maintenance and Weed Control

Tufted hair grass has a slow juvenile development. Therefore, crop residues should be removed quickly. It is recommended that the stubbles are cut once more. Before sowing, only contact herbicides should be used, up to three days before the beginning of sowing. Hormone-type and broad-spectrum herbicides should generally not be used until the species reaches the three leaf stage.

In order to obtain optimal development up to autumn of the seeding year, weed control has to start in time. This species is quite robust and resistant. However, specific control of weed grasses is not possible, as it is the case with the gender Festuca. The use of curry combs during mechanical weed control is well tolerated. Comparative experiments show that there will be higher yields in the second harvesting year if the board of the thresher is kept very high at the first harvest. This method also avoids the leaves being cut. Leaves die off over winter. Legal regulations permit the cutting or burning of the biomass in early spring.

Harvest and Yields

Resistance to lodging: high.

Shattering tendency: intermediate.

Ripeness: the flower spikes change colour from greenish-yellow to bright yellow or straw-coloured. The seeds will shed if touched only lightly. As with most species common to the habitat, great attention should be paid to specifiying the optimal harvest time. Daily or less frequent control of the population can help in deciding.

At the time of complete ripeness almost all species are very sensitive to wind and heavy rainfall, and it is possible to lose up to 70 percent of the crop within a few minutes.

Ripening period: depending on climate and location, ripening takes place between the beginning and the middle of July.

Harvesting technique: for threshing it is recommended that the rotational speed of the drum be slightly increased and the settings of the concave narrowed substantially. The seeds have sufficient seed flow and can be cleaned easily.

Crop yields: yields can fluctuate considerably from year to year. As a rule, higher yields can be expected in the second harvesting year. Average yield is around 160 kg/ha, where as single yields can be noticeably higher.

FESTUCA NIGRESCENS

Alpine chewing's fescue

In the Alps this species can be found in montane,

subalpine and alpine areas. Indifferent to soil type, it can therefore be found on almost any kind of rock. However, it prefers predominantly dry and sunny locations. In moist meadows and pastures it grows less. In the Alps it flourishes at altitudes higher than 2 600 m. All-purpose high-quality forage in montane to alpine regions; resistant to high nutrients, grazing and cutting.

Seeds

Fruit of the husk 3.5-6 mm long, 0.5-1 mm wide, 0.5-0.8 mm thick, oblong, acuminate, greyish-yellow to brown, often stained red. Awn up to 2.5 mm long. Lemma vaulted, keeled because of midvein, marginal veins indistinct, short hairs on margin and midvein, palea in a hollow, fruit dark brown to red. Keels apically dentate, short.

Rachis 1.2 mm long, straight, slightly expanding, distal surface broadening into a head shape. Fruit connate, 3.0-4.0 mm long up to 1 mm wide, 0.5 mm thick, apically lanceolate and round, base acuminate, ventrally grooved, dorsally ellipsoid, red to brown.

Thousand seed weight: 0.8-1.2 g.

Chewing's fescue has no special soil requirements. It roots deeply and therefore shows no sensitivity to dryness. It prefers intermediate to highly nutritious soil. The pH should be between 5.5 and 6.8. Moist or waterlogged soil, peat, ground that tends to grow weeds and light, dry soil should be avoided. Compared to other alpine plants it is more competitive towards weeds. Nonetheless, locations with a high presence of quack grass, meadow grass and crab grass should be avoided. Cultivation of seeds is possible up to an altitude of 1 000 m.

Cultivation

Open sowing is possible if carried out until the beginning of July or (in mild climates) middle of July. Seeding below a cover crop in autumn is possible with thin cover crop populations. Cultivation under winter barley is therefore not recommended cover crops in spring are generally unproblematic. Seeding can be carried out underneath various summer cereals and linseed.

Juvenile development is faster in comparison to most other site-specific grasses; nevertheless, seeding should be done immediately after the seeding of the cover crop.

Seed depth: no more than 0.5 cm.

Seed rate: 6-8 kg, depending on experience and cover crop.

Row spacing: broadcast sowing or row spacing of 12-25 cm, depending on the construction of the seeder. The aim should be to grow well-developed and strong single plants.

Fertilisation

Chewing's fescue is generally undemanding. To obtain a satisfactory yield, nutrient supply should be high.

Phosphorus and potassium: on soils with an intermediate phosphorus and potassium content, fertilisation with manure (15-20 tonnes/ha) is sufficient. For mineral fertilisation 70 kg/ha P_2O_5 and 80-120 kg/ha K_2O are recommended.

Nitrogen: a sufficient supply of nitrogen in autumn guarantees satisfactory tillering. An excessive supply of nitrogen in late spring may cause decreased formation of spermatophores. However, a certain amount of nitrogen is necessary for the formation of seeds and 70-100 kg/ha of Ntotal should be split and applied in autumn and early spring.

Maintenance and Weed Control

Guidelines for the gender Festuca generally apply. However, since cespitose stands form only a few seeds, densely growing turfs should be thinned in late summer before the second harvest year. Various machines (e.g. disc harrows and grubbers) can be used to stimulate tillering in order to obtain a generous crop. In well-developed populations, cutting is recommended in the first autumn, but is not necessary in the following one.

Optimal Ripeness for Harvesting

Removing biomass is usually not necessary. Weed control can be undertaken mechanically. For stands with 25-cm row spacing, a hoe or a brush can be used. Application of fungicides against rust in late summer of the first harvest year is seldom required. As for all fescues, selective control of weed grasses is possible. The most intrusive grasses can be eradicated sufficiently without affecting the crop.

Harvest and Yields

Resistance to lodging: high (in small populations) to intermediate (in wellmaintained seed populations with high crop yields).

Shattering tendency: low to intermediate. Where there is a satisfactory nutrient supply, the whole stand lodges after flowering, which strongly reduces the shattering tendency.

Ripeness: because of the low shattering tendency, fixing a time for flailing is not difficult. Culms change colour to yellow and seeds become firm and brown, often stained red. Threshing time is indicated by the number of shattering seeds when a bunch of spikes is beaten against the palm of the hand.

Ripening period: ripeness is to be expected between the middle and end of June, depending on the origin of the plant material, climate and soil conditions.

Harvesting techniques: threshing has proved unproblematic. Seeds have a good seed flow and can be cleaned easily. As for all seeds, immediate and gentle drying is necessary.

Crop yields: between 600 and 1 000 kg/ha in the first year and 400 and 700 kg/ha in the second year, depending on practical experience.

Koeleria Pyramidata

June grass, crested hair grass

Very often on oligotrophic grassland and arid pastures as well as meadows, grass heaths and exposed hills, roadsides, in low-density stands of pine and oak forests, edges of forests and scrubs. On moderately arid, poor, usually limy, neutral, humus loess, loam or clay soil, seldom on sand. Indicates moderate aridity and poorness. Disappears with excessive fertilisation and irrigation. From lowlands to subalpine zones, in the Alps up to 2 300 m above sea level. All-purpose species on poor and arid slopes and hillsides up to the subalpine and alpine zones.

Seeds

Glumes dissimilar, keeled, smooth and glabrous or scabrous; patulous, short cilia on midvein; membranous, broad, finely membranous margins, lower glume one-veined, 4-5 mm long, lanceolate when viewed laterally, pointed, upper glume three-veined, 5-6 mm long, broadly lanceolate, pointed.

Lemma three-veined, 4-6 mm long, lanceolate, acuminate or ending in a short awn tip, glabrous, shortly pilose or seldom ciliated, membranous, with finely membranous margins. Palea two-veined, 3.5-5.5 mm long, lanceolate, apically notched, with pointed lateral lobes. Anther 2-2.5 mm long. Fruit 2.5-3 mm long, glabrous, laterally contracted, outline narrowly elliptic.

Thousand seed weight: 0.4-0.5 g.

Crested hair grass has no special soil requirements and is therefore not sensitive to aridity. It prefers an intermediate nutrient supply. The pH should be between 6.0 and 7.5. Moist or waterlogged soil, peat, ground with high weed infestation and light, dry, sandy soil should be avoided. Competition with weeds is low and juvenile development slow. Locations with a high presence of Agropyron repens, Digitaria sanguinalis, Poa annua and Poa trivialis are to be avoided.

Cultivation

Open sowing is possible if carried out early in the year (until the middle of June). A high expenditure for weed control is to be expected. Because of the plant's slow juvenile development and low competition, seeding below cover crops should be avoided in autumn.

Spring seeding under summer cereals and linseed is unproblematic. Because of photosensitivity and slow juvenile development it is important to have a thin cover crop population. Seeding should be done immediately after the seeding of the cover crop.

Seed depth: no more than 0.5 cm.

Seed rate: 7-9 kg/ha.

Row spacing: 12-15 cm. Broadcast seeding is possible.

Fertilisation

Phosphorus and potassium: moderate fertilisation with phosphorus (40-60 kg/ha P_2O_5) and potassium (80-100 kg/ha K_2O) has proved to be worthwhile although crested hair grass has low nutrient requirements. Timely fertilisation with liquid or solid manure in late summer or early autumn stimulates the development of single plants until the resting season. Application of liquid manure in spring should be avoided because it may lead to the mobilisation of nitrogen in disadvantageous periods of the growing season.

Nitrogen: total requirements for seed development are 60-70 kg/ha. In spring, nitrogen should only be applied as mineral fertiliser. The amount is 30-50 kg/ha, depending on climate and weather.

Maintenance and Weed Control

Because of the plant's low growth height and the associated danger of being stalled by crop residues, it is important that the stubbles are cut and the straw removed immediately after harvesting the cover crop. As for most grasses the following guidelines apply. Before sowing only contact herbicides should be used, up to three days before the beginning of sowing. Hormone-type and broad-spectrum herbicides should generally not be used until the species has reached the three leaf stage.

Even with good maintenance after harvesting the cover crop it is not possible to obtain a compact stand until autumn. The low growth height of this species means considerable effort in maintenance and weed control. Selective grass weed control is not feasible. Where weed grasses prevail, a specific application of Ethofumesate will control annual meadow grass in particular. However, an expert should be responsible for evaluation and composition. National guidelines concerning the use of herbicides should be followed.

Selective control with wick and total herbicides (Glyphosate) is possible during most periods of the growing season because of the different growth heights of wavy hair grass and other grasses.

Harvest and Yields

Resistance to lodging: high, in excellently maintained populations intermediate.

Shattering tendency: intermediate.

Ripeness: panicles and culms change colour to light tawny. Seeds adhere well to the panicle after ripeness.

To check ripeness: if a bunch of panicles is grasped, the seeds will fall off easily if ripe.

Flowering Single Plant

Ripening period: last decade of June to first decade of July.

Harvesting techniques: crested hair grass can easily be threshed from the root. The rotational speed of the threshing drum should be set low to avoid damage or dehusking of seeds. The air supply should be reduced. Seed flow is sufficient; seeds as well as the chaff in the sieve should be checked several times while threshing. Settings of the threshing concave should be close.

Cleaning of seeds is usually unproblematic.

Crop yields: yields can vary greatly (80-150 kg). Very little experience has been gained regarding the propagation of this species. With experience, 150 kg and more can be expected. For optimally maintained populations three harvesting years are possible

LEONTODON HISPIDUS S.L.

Rough hawkbit

European species that is extremely polymorphic, common and occurs in the Alps, from the foothills to the upper alpine zone. On (alpine) meadows and pastures that are moist to slightly alternating moist and dry, at low-density forest sites, on talus sites and sometimes on rocks.

Multifunctional use for restoration at all altitudes, good forage, filling gaps well. The glabrous subspecies Leontodon hispidus ssp. hastilis grows in the subalpine and alpine zones.

Seeds

Achene 6-7.5 mm long, 0.7-0.9 mm wide and 0.5-0.7 mm thick; oblong-bacillary, only slightly tapered at both margins, slightly angularly compressed, with fine lateral grooves, highly cross-rugose. Pappus brownish-white to yellowish, outer hairs are simple bristles, inner ones lanceolate leaflets fading to pinnate hairs, up to 10 mm long (disappearing when cleaned). Surface auburn to dark brown, slightly glossy.

Thousand seed weight: 1.3-1.5 g.

Rough hawkbit has no special soil requirements. Because of its deep roots it is minimally sensitive to drought. It prefers an intermediate to good nutrient supply. The pH should range between 5.5 and 7.5. Soils that are wet, waterlogged, or tend to a high weed infestation should be avoided, as well as soils that are rich in peat, light, or dry.

Cultivation

Open sowing is unproblematic, if sowing is carried out by the beginning of July at the latest (at mild locations by the middle of July). In the latter case irrigation of the crop is recommended. To date no experience has been gained with cover crops.

Seed rate: 10-12 kg/ha are necessary for a satisfactory population density (if the seed quality is good).

Row spacing: 15-25 cm. Depending on the construction of the seeder and the method of weed control there should be either dense populations or sufficient row spacing (for mechanical and/or chemical weed control). Broadcast sowing is generally possible.

Fertilisation

Phosphorus and potassium: on soils with an intermediate phosphorus and potassium content, basic fertilisation with solid or liquid manure is sufficient in autumn. If the soil content is on high level, fertilisation with 40 kg/ha P_2O_5 and 60-80 kg/ha K_2O is considered sufficient, depending on the yield.

Nitrogen: at the sowing stage, fertilisation with about 30 kg/ha of N-total is recommended in order to obtain the best development up to autumn. An application of 60-80 kg/ha of N-total is necessary per year, to be added in autumn, early spring and after the first harvest. If fertilisation in spring is carried out too late, the principal outcome will be stimulation of the leaf mass.

Maintenance and Weed Control

In early autumn before the crop year, cutting makes sense if the populations are densely grown. If competition with weeds is low, mechanical weed control is sufficient. Rough hawkbit is not sensitive to the use of curry combs. As for all herbs, the control of weed grasses is relatively easy. On the other hand, herbicide control of dicotyledonous weed species is problematical. It is possible to control weeds, mainly in summer and autumn of the seeding year, as well as in spring of the harvesting year, by wiping wick and total herbicides (Glyphosate), and taking advantage of the differences in growth heights.

Pests and diseases: leaf diseases caused by fungal decay are possible, but do not generally lead to losses in crop yields. At the first harvest, larvae of flies that mine in the flowering heads can cause significant problems. Ongoing control of the flowering heads is recommended. If there is decay, using a systemic insecticide against mining and sucking pests (such as 0.5 l/ha Deltamethrin) is imperative, in order to make the first harvest possible.

Harvest and Yields

Resistance to lodging: high.

Shattering tendency: high.

Ripeness: after flowering, the flowering heads close and open again after three to five days.

The seeds adhere to the flowering head quite firmly for one or two more days, after which they shed easily, even at a puff. Consequently, the seeds should be threshed, when only 40-50 percent of the flowering heads are open.

Ripening period: first harvest during the first and second decade of June; second harvest between the first decade of August and the beginning of September, depending on the weather.

Harvesting technique: direct threshing is possible. If ripening is unequal, very large sieves, a low rotational speed, and a wide gap between threshing drum and concave are recommended. In this way the withered but still closed flowering heads can also be harvested. After one or two days of cool air ventilation, seeds enter an after-ripening stage, which leads to a belated burst of the flowering heads.

As an alternative, harvesting is possible using suction devices whereby the ripe seeds are harvested without destroying the mother plants. Because only completely open flowering heads ensure an efficient harvest, harvesting should take place on warm, sunny days with low air humidity. The first harvest is recommended when about 50 percent of the flowering heads are open. After most of the flowering heads have been harvested, cutting is recommended, in order to stimulate seed formation once again. Altogether up to eight harvests per season can be expected, using this method.

Flowering Plants

Crop yields: because of lack of experience regarding the cultivation of rough hawkbit, yield forecasts are difficult. Yields range between 60 and 180 kg/ha, where the amount of harvests per year are combined. Optimal maintenance allows up to three harvesting years.

7

Production of Forest-based Medicinal Plants

Rural people in the Asian countries are familiar with the medicinal properties of plants growing close to their homes, in the open fields, margin of the water bodies, waste lands and the near by forests. The herbal doctors in the villages are well acquainted with various plants they need for different ailments. The knowledge about these plants and the various formulations made were usually kept as family or professional secrets. They are passed on from the parent to the offspring orally and through demonstrations.

The bio-diversity of medicinal plants in India is associated with the rich cultural diversity related to traditional system of medicine. Besides the known reasons for the loss of plant bio-diversity, the disintegration of cultural diversity is also responsible for the destruction of medicinal plants. According to World Health Organisation (WHO) more than 80 per cent of the world's population in the developing countries depend on traditional medicine fcr their primary health care.

Over 1.3 billion people in the world can hardly afford to spend any money on modern medicine and therefore have to resort to local medicinal plants for their health needs. Out of the 350,000 higher plants identified so far about 35,000 (the estimates vary) species have at one time or other used by some people or cultures for medicinal purpose. Presently there is a resurgence of natural product-based industries and pharmaceutical products because of the interest created in the developed countries for traditional medicine and natural products. Consequently the traditional medicine has now becoming more costly and unaffordable to the poor. The revitalisation of the local health tradition would therefore lead to the conservation of rich traditional knowledge and the species used by the local people. This also ensures availability of effective and cheap health care needs to the local poor.

Tropical forests are the storehouse of bio-diversity and natural resources. Rainforest contains no less than 60 per cent of all higher plant species known on earth and they provide everything needed for human survival. Through many years of evolution plants have developed large numbers of chemical substance to defend themselves against pest and pathogenic diseases. These chemical substances, at least some of them, can act within the human body against micro-organisms and other causes of diseases. They represent important sources of natural drugs. They are highly complex and their molecular structures surpass the imagination of the chemists and cannot easily be reproduced in the laboratory.

Biodiversity and Medicinal Plants

Rainforests contain no less than 60% of all higher plant species known on earth and they provide all that is needed for human survival, including remedies against disease. Through evolution, plants have developed large numbers of chemical substances to defend themselves against insect pests and fungal and other pathogenic diseases. Some of these agents can also act within the human body against microorganisms and other causes of disease, and represent an important source of natural drugs. Their highly complex molecular structures often surpass the imagination of the chemist and cannot easily be reproduced in the laboratory.

More than 35,000 plant species are being used in various human cultures around the world for medical purposes and many of them are subjected to uncontrolled local and external trade. So far, natural products from fewer than 40 tropical species have been incorporated into modern medicine and only a fraction of the tropical flora has been thoroughly analysed for their pharmacological activity. Therefore, the annual extinction rate of an estimated 3,000 plant species is a matter of great concern as it could imply the loss of a potential drug against an incurable condition, such as dementia, cancer, influenza or AIDS.

Traditional Health Care

Health care and botany have evolved as inseparable domains of human activity: the medicine man (shaman) is often regarded as the first botanical professional in human history. Whereas western medicine, as taught in most medical schools around the world, has largely switched from natural to manufactured drugs, plant products are still of paramount importance in traditional health care systems of developing countries.

In traditional therapies of certain indigenous communities, herbs are administered along with chants, dance and spiritual ceremonies to expel bad spirits and to help reharmonising the sick person with his or her environment. Plants, however, also serve a less metaphysical role, as anticonceptives in indigenous birth control procedures or to counteract tangible pathogenics such as fungi and parasites (e.g. worms, malaria).

In developing countries, medicine men and women are particularly knowledgeable about the recognition and treatment of common diseases. In Amazonia, at least 1,300 plant species are being used as medicines, poisons or narcotics. Traditional healers are also skilled botanists and have a great talent for locating the requisite plant from the green vastness that makes up their natural pharmacy.

In Latin America and Africa, this knowledge has largely remained undocumented and is handed down orally from father to son or from mother to daughter. Today's younger generations often have very different ambitions and, therefore, these traditional skills are doomed to get lost even faster than the plants themselves. This is why ethnobotanists compare the death of a shaman to the loss of a national library and invest much effort in assembling this knowledge as written accounts.

In recent years, traditional healing in distant forest areas has come under pressure from novel diseases such as influenza and tuberculosis, that have often revealed the superiority of 'white man's capsules'. This course of events has greatly affected the prestige of the local healers and has also opened the market for expensive and less necessary western drugs. Apart from the heavy drain imposed on foreign exchange reserves by these imports, the existing available and often equally effective traditional equivalents have been forced into disuse and oblivion.

Although western medicine has been integrated to some extent with ancient health care systems in Asia, it has become the dominant method in most larger hospitals around the world. In non-hospital care in most developing countries, traditional and modern systems operate independently without a clear hierarchy, whereas in rural areas only traditional healing and herbal self-care may be at hand. In these countries, there is a great demand for medicinal plants that often come from the forests.

In the future, the use of these plants can be expected to increase further due to population growth and the increasing importance being attached to traditional health care by the World Health Organisation (WHO). WHO's 'Health for All by the Year 2000' campaign emphasises the urgent need for the conservation of medicinal plants. This same initiative has launched a first step toward a more rational use of herbal medicines in the Carribean, following a recent pharmacological evaluation of their effectiveness and safety.

In Industrialised Societies

In the second half of this century, 'chemical' drugs have largely replaced plant products in mainstream medicine. This development is in line with the prevailing concept of disease, the belief in human-initiated progress, and the quest for pure therapeutic substances that contain no more than one active principle.

On closer analysis, however, over 25% of all prescription drugs in the Organisation for Economic Co-operation and Development (OECD) countries, and up to 60% of those in Eastern Europe, prove to consist of unmodified or slightly altered higher plant products. They embrace such important therapeutic categories as anticonceptives, steroids (e.g. prednisone) and muscle relaxants for anaesthesia and abdominal surgery (all made from the wild yam); quinine and artemisinin against malaria; digitalis derivatives for heart failure; and the anticancer drugs vinblastin / vincristin, etoposide and taxol. These agents cannot be (fully) synthesised in a cost-effective manner.

Therefore, their production requires reliable supplies of plant material, either from cultivation or from the wild. Such is the case for the wild yam *Dioscorea composita* (Hemsl.), which cannot be cultivated and is exported from

Mexico and other countries in quantities of hundreds of tons.

These few examples should make one realise how modern drug delivery depends on the continuing availability of raw materials and how vulnerable it is to the exhaustion of natural resources. This awareness is even more pertinent to clients of health products stores and herbalists, since natural non-prescription formulae may contain rare and even endangered wild plants that are regarded, often unjustly, as more powerful than their cultivated analogues. So far, only three species of medicinal plants have been listed by CITES, the Convention on International Trade in Endangered Species of Wild Fauna and Flora.

Natural Drugs and Medicinal Plants

Natural drugs and medicinal plants, along with other non-timber forest products, already yield important economic returns. They compare favourably in monetary terms with logging and cash cropping, and contribute in providing better prospects for sustainable forest use.

The pharmaceutical industry has recently rediscovered the tropical rainforests as an unmatched source of chemicals with potential for new drug development, which promises additional revenues. Thousands of plant extracts of all continents are being screened for activity against HIV and cancer in the laboratories of the U.S. National Cancer Institute.

For example, Merck Sharpe & Dome, a New Jersey based pharmaceutical company, has paid US$1 million for research rights in Costa Rica and has agreed to contribute 25% of profits made from Costa Rican plants to rainforest conservation in Costa Rica. Shaman Pharmaceuticals Inc., a pioneer in ethno-directed natural product research since

1989, considers indigenous people as partners and only collects plant samples on the indication of a shaman.

This approach appears to be more effective than random collection methods and has already resulted in the discovery of three novel drugs. Before long, one of these drugs, SP 303, will become available as a neat monocomponent drug against herpes simplex and secretory diarrhea. SP 303, soon to be known as Provir and Virend, originates from an ordinary weed in Peru that can be harvested in a sustainable way. The expected market value may be hundreds of millions of U.S. dollars. Part of the profits will be returned to the Indian communities in the form of support for cultural and health care projects and nurseries for endangered medicinal plants. As a further consequence of this development, the prestige of the shaman will be upgraded, also in western eyes.

Health Care and Ethnobotanical Research in Developing Countries

According to recent estimates by the World Health Organisation, more than 3.5 billion people in the developing world rely on plants as components of their primary health care. Just as many Europeans know of the use of *Aloe vera* [Aloaceae] to treat burns, many indigenous peoples know of some common plants that have medicinal uses. Ethnobotanical research should not be limited to discovering new pharmaceuticals for Westerners; it can also be of some benefit to peoples in developing countries.

An increasing number of nations, including China, Mexico, Nigeria, and Thailand, have decided to integrate traditional medicine into their primary health systems. In these systems, ethnobotanical research plays a crucial role in documenting the traditional health care practices of the country. Medicinal plant lore often recedes or completely vanishes in the wake of rapid Westernisation.

In some countries, careful ethnobotanical studies have become invaluable records of ancestral ways. In areas where the people are moving away from traditional lifestyles, particularly in rapidly growing urban populations, careful ethnobotanical documentation can provide the needed foundation for educational programmes. Workers at Mahidol University in Bangkok, for example, have prepared a series of slide presentations and pamphlets to teach schoolchildren about traditional Thai uses of plants.

Ethnobotanical research can also help in the discovery of crude drugs. Only pure compounds with known structures and pharmacological activities are permissible as drugs in Western medicine, but in many developing countries the price of such pure substances puts them beyond the reach of all but the affluent.

Careful clinical studies can document the safety and efficacy of crude extracts or tinctures of plants that can be dispensed at far less cost. Carefully designed clinical trials of crude botanical drugs have been conducted in Mexico and Thailand.

An area of ethnobotanical drug discovery that has yet to be developed is that of "gray pharmaceuticals"-drugs of proven safety and efficacy that are not marketable in the Western world. Decisions concerning marketability in the Western pharmaceutical industry are not driven solely by proof of safety and efficacy. To be marketable, a drug candidate must affect only one point on a biochemical pathway: compounds that affect multiple points of the same pathway are unlikely to be marketed because only "magic bullets" (single-activity drugs) are viable in today's legal and economic environment.

Drug candidates must also show superiority over competing drugs in the same market. Thus some plant-derived drugs that are not marketable as Western pharmaceuticals may still be acceptable in the country of their origin, particularly if they can be produced cheaply. The transfer of information (sometimes costing millions of dollars) concerning the safety and efficacy of such gray pharmaceuticals from Western firms to developing countries, should be encouraged, along with the appropriate patent rights and technologies to enable the developing countries to produce plant-based medicines.

Intellectual property Rights

Historically, the intellectual property rights of indigenous peoples have not been recognised. The use of information supplied by indigenous peoples in the discovery of commercially marketable pharmaceuticals raises the question of those people's intellectual property rights and the ownership of biodiversity.

We believe that indigenous peoples are entitled to the same intellectual property rights enjoyed by other investigators. Yet in many cultures, the preservation of important habitats is equally urgent. In Samoa, four village-owned and -managed reserves totalling 50,000 acres, beginning with the Falealupo Rain Forest Reserve (where the tree that produces prostratin was first collected), have been created with donated funds, and in Belize, the world's first extractive reserve for medicinal plants has been created on 6000 acres of tropical rain forest by the local government working with the association of traditional healers with significant international support.

This effort seeks to demonstrate that conservation and the use of forests as sources of locally consumed medicines are compatible objectives. Elsewhere, as in India, medicinal plant reserves are being established to ensure a continued supply of plants for traditional health care practitioners and their patients.

Cash disbursement of royalty income most closely approaches the Western concept of equity, but this approach fails with peoples who have no monetary system. For many indigenous peoples, the right to live unmolested and undisturbed on their ancestral lands is the greatest value. Establishment of nature preserves that protect both biodiversity and indigenous cultures is of tremendous importance to indigenous peoples.

Original Perspectives on Conservation

For hundreds of years Savaii Island, like the rest of Samoa, was, in Somerset Maugham's words, "lovely, lost and half a world away." Increasing prices of rain forest timber in the early 1970s ended that isolation. An American timber firm built a large sawmill in Asau, Savaii. Lack of experience with both the forest and the culture of Samoa eventually caused the firm to jettison the project.

The sawmill, however, remained in operation under various owners, and the rain forests of Savaii continued to disappear. Now more than 80 percent of Savaii's forests are gone forever: only two large tracts of lowland rain forest remain, one of them the Tafua rain forest. Because of its proximity to the wharf, the Tafua forest offered the most lucrative logging opportunity in all of Samoa. The logging companies faced only one small problem: the paramount orator and chief, Ulu Taufa'asisina.

Samoans are gentle but determined people; but even by Samoan standards, Ulu was resolute. Although Tafua is a very poor village with no running water, electricity, or graded roads and few sources of cash from the loggers, not a single tree could be cut. The villagers begged Ulu to accept the logging companies' generous offers. How else could the village pay for a decent school for their children or a clinic for their sick and elderly? The loggers might even hire some of the villagers to work for them. Ulu's stance mystified the logging companies' representatives too. They were offering the village what would probably be its only chance for economic development.

The loggers failed to realise that no inducement could ever persuade Ulu Taufa'asisina to allow logging. When his father lay dying, Ulu had promised to honour his last wish: Ulu had pledged to protect the rain forest with his life. Ulu Taufa'asisina has paid a price for conservation that few individuals in industrialised countries can comprehend: he has knowingly condemned his family, friends, and village to poverty rather than accept money from loggers. "Five times the logging companies have been here asking for our forest," Ulu explains.

> I was deeply depressed because they put a lot of pressure on all of us, persuading the people of my village to sell the forest for a few dollars. I resisted, because I love my land and the land more than the money.
>
> The land is our lives. The land is also our mother. The land is sacred. I believe that the land has provided the culture, the food, the water, and the other things essential for my people. I deeply respect the honour that has been given to me, as chief orator, to become a caretaker of our beloved land.
>
> My forefathers had a dream. They had a dream that one day the land and the rain forest would be saved for eternity. They had a dream that the land and the sea would forever be well looked after, and not destroyed and distributed to other people. I share that dream. I believe that we can become masters of our destiny if we take care of our environment.

That destiny encompasses tooth-billed pigeons, flying foxes, and dolphins-and relentless economic hardship. But Ulu Taufa'asisina, like many other indigenous leaders, sees conservation in terms that transcend economic or political issues.

In the Western tradition, natural resources are property and therefore subject to either private or government ownership. Thus Western conservation has its roots in the pragmatic use of property; according to this viewpoint, no action should be taken that decreases the value of the resource for the long term. Many indigenous cultures, in contrast, perceive the earth as existing not in the realm of the profane, but in the realm of the sacred, a world view that distinguishes them from many Western traditions. Indigenous legends emphasise the need to protect the earth not because it is useful to humans, but because it is sacred. The perception of conservation as a religious duty, of course, also serves ecological and cultural purposes.

Though they start from very different assumptions, both Western conservationists and many indigenous peoples recognise the need to protect vanishing natural habitats. When Maori elders became concerned about the loss of native plants used in weaving, for example, they organised a hui, or a traditional conference, with the New Zealand Division of Scientific and Industrial Research (DSIR). They invited both scientists and traditional leaders to discuss conservation strategies.

Such collaborations, although complicated by cultural differences, have provided strong support for three positions advocated by indigenous peoples: that all forest plants have a purpose and value; that the true economic (let alone cultural and spiritual) values of rain forests and native habitats have scarcely been considered and are vastly underestimated; and that entire cultures and ways of life will disappear if rain forests are destroyed. Recent ethnobotanical studies provide evidence that substantiates these indigenous views.

Ethnobotany in South America

Forest-dwelling peoples often claim that most, and perhaps all, plants in their environment have a use. Ethnobotanist Brian Boom of The New York Botanical Garden used some pioneering plant census techniques to test that hypothesis. Working for an extended period in the Bolivian Amazon, Boom found that the Chácabo Indians knew of 360 species of vascular plants in the forest surrounding their village of Alto Ivón and that they had uses for 305 of them.

They collected Brazil nuts (*Bertholletia excelsa* [Lecythidaceae]) for their own consumption and for sale, for instance, and used *Anthurium gracile* [Araceae] to cure appendicitis. Boom then surveyed a 1-hectare plot in the tropical forest and found that 82 percent of the tree species growing there had uses known to the Chácabo. When he measured the densities of plants in the plot, Boom found that the Chácabo used 95 percent of the individual trees for some purpose.

The percentage of tree species put to use by the Ka'apor was found to be 76.8 percent, by the Tembé 61.3 percent, and by the Panare 48.6 percent. Although these findings do not prove that every forest plant has a use, they do confirm the local people's claim that the forest plants have far more uses than Western investigators have realised. Certain plant families are so important in these Neotropical areas that conserving them is essential if people are to continue to depend on the forest for their sustenance. Among these valuable plant families are the palm family [Arecaceae], the Brazil nut family [Lecythidaceae], a tropical relative of the rose family [Chrysobalanaceae], and a family that includes the hallucinogenic *capi* vine, Malpighiaceae.

Working in Tambopata, Peru, with mestizo people, Oliver Phillips and the late Alwyn Gentry of the Missouri

Botanical Garden employed an even more detailed quantitative technique to inventory the plant families used for construction, in commerce, for food, for technology, and for medicine. Their interviews with 29 field guides yielded a total of 1885 reported uses for the 605 tagged plants in their plots. When they compared data to determine whether the age of the person interviewed affected his or her knowledge of plant use, they found that the bulk of the information in some categories, such as medicinal plant lore, was held by older people.

These are the people, who should be the main focus of ethnobotanical studies and conservation efforts. Through their use of statistical tools, they substantiated the intuitive judgements of many other workers, who perceived that the long chain of oral ethnomedical tradition was coming unravelled in the current generation. Once investigators can identify the best sources of ethnobotanical information in a community or indigenous society, both local people and ethnoscientists can make more efficient efforts to conserve such information.

Forests are more than Timber

Historically rain forests have been cut down because the simplest and quickest way to convert them into cash is to harvest the timber, burn down all that remains, and plant and annual crop for a few seasons, until much of the soil's nutrients are leached out. Since most of the nutrients in the tropical rain forest are found in plant material rather than in the soil, large-scale removal of that living material (called biomass) prevents rain forests from growing back. By using the tools of economists to analyse the value of land under various uses, ethnobotanists have found that in some areas there are viable alternatives to clear-cutting.

Researchers studying the use of non-timber resources from forests in Brazil and Peru have concluded that non-wood forest products "yield higher net revenues per hectare than timber, but they can also be harvested with considerably less damage to the ecosystem. Without question, the sustainable exploitation of non-wood forest resources represent the most immediate and profitable method for integrating the use and conservation of Amazonian forests."

In a similar study, Balick and Mendelson valued the native medicinal plant species taken by the local people from a forest in Belize. From two separate 1-hectare plots of 30- and 50-year-old forest, respectively, total biomass of 308.6 and 1433.6 kilograms (dry weight) of plant material for medicines was collected. It was suggested that harvesting the medicinal plants, from a hectare of forest would yield the collector $564 and $3054 in the local markets, respectively, for the two plots, after the costs of harvesting, processing, and shipping were subtracted. For the 30-year rotation, a present value of $726 per hectare was calculated for the medicine, and for the 50-year rotation a present value of $3327 per hectare was calculated.

This study fostered a greater understanding of the value of the tropical forest to the local inhabitants and their economy. It ultimately led to the development of several industries based on the extraction of medicinal plants from the forest for processing into tinctures, extracts, and salves. Today local Belizean brands of traditional medicines-Agapi, Rainforest Remedies, Rainforest Rescue, Triple Moon-all help to generate employment for many local people.

Additional studies to establish the net present value of the tropical forest of the Neotropics have confirmed the relatively high value (often several thousands of dollars) of the products that can be harvested on a hectare in areas where land is now priced in the hundreds of dollars or less. Critics of this method of establishing the value of forests point out that the land must be near to a market or to a distribution channel in order for the economic benefit to be realised.

They state that there is probably a finite market for the commodities produced under these management schemes. Both points have their validity, but nevertheless, it is clear that in areas that have been intensively studied the harvesting of non-timber forest products has increased the income levels of local people and has stimulated the development of new industries with a local value-added component that increases returns to the region or country of origin. Such studies also tend to confirm indigenous beliefs that tropical forests, if properly managed, have far more value than as mere sources of lumber and wood pulp.

"Green" industries now promote the sale of rain forest products, such as the buttons fashioned from palm seeds that adorn garments made from Paris to Hong Kong, ice creams flavoured with exotic nuts and fruits, and rare tropical essences in perfumes, shampoos, and body creams. We can wash with soap made from tropical oils and nectars, eat cereals based on grains that once sustained the Aztecs, and drive to work on tyres manufactured from wild rubber harvested in the Amazon Basin. Like other suggested solutions to the dilemmas posed by deforestation and economic development, green marketing is hardly a panacea. The continued use of these products depends on the reliability of their supply, markets, and distribution. Key to maintaining supply is the issue of sustainable resource production.

Goods from the Woods

Of greatest concern in the development of products based on tropical forest species (usually known as non-timber forest products, or NTFPs) is our ability to ensure sustainability. But nowadays, the concept of sustainability is used in a rather cavalier fashion. In truth, we know very little about the sustainability of any production, especially of the products from tropical ecosystems.

Charles Peters of The New York Botanical Garden has undertaken many detailed studies of tropical forest trees in efforts to determine the level of sustainable production or harvest of each species. According to Peters, "a sustainable system for exploiting non-timber forest resources is one in which fruits, nuts, latexes, and other products can be harvested indefinitely from a limited area of forest with negligible impact on the structure and dynamics of the plant populations being exploited.

" A plant such as *Brosimum alicastrum* [Moraceae], a tree found in Central and South America that is exploited for its protein-rich fruits, needs to produce over 1.5 million seeds to ensure that *one* tree will live long enough to reproduce. If most of the fruits produced by this species were to be harvested rather than left to grow in the forest, the population would become extinct within one generation.

Too little is known about the levels of sustainable harvest of many of the internationally important NTFPs, including the Brazil nut. Some 200,000 people harvest the Brazil nut from the 20 million hectares of Amazonian forest where it grows; annually they produce around 42,000 metric tons for the commercial trade, valued at approximately $35 million, or 1.5 percent of the total international nut trade.

The harvest of this nut is one of the largest sources of cash income for many of these people, and reduction in government subsidies for other NTFPs such as rubber, which grows in the same areas, has led people to harvest increased quantities of Brazil nuts. Quite simply, the mature, seed-producing trees that are the backbone of the population will die and not be replaced, and the resource base on which these industries are built will disappear.

Charles Peters suggests six steps for exploiting NTFPs in a sustainable fashion. First, the species to be exploited should be carefully selected, after such factors as the ease of harvesting and resilience of natural populations to disturbance are considered. A tree valued for its roots will be harder to harvest than one valued for its fruits, and the harvest of a species that produces fruits in massive quantities at one time of year will be easier to manage than the harvest of a species that produces fruits sporadically throughout the year.

Once the species has been decided upon, a forest inventory should be undertaken to learn where the resource is found in greatest abundance and the number of productive plants per hectare. Investigators should then estimate the quantity of the resource produced by the species in its various habitats and by trees in all size classifications, to determine which trees in which habitat it is best to harvest.

When these three steps have been taken, the harvesting of the resource can begin, but the careful measurement should continue. The status of the population should be monitored for signs that the forest is being overharvested. People should examine the status of adult trees periodically to determine whether the flowers are being pollinated, whether large numbers of fruits are being consumed by predators, and so on. If problems arise, the harvest should be adjusted to keep its level below the rate that would threaten sustainability.

When necessary, people may replant areas that do not seem to be regenerating, clean out competitive species, or open up the forest canopies to allow more light to reach the young trees and thus speed up their regrowth. The precise measurements that Peters recommends are expensive and time-consuming, and very few species have been studied from this perspective. However, plant populations may be threatened if harvests are determined by the demands of the marketplace rather than the needs of the ecosystem.

As Peters notes, "nature does not offer a free lunch." In our enthusiasm to support conservation of the natural world by focusing on its usefulness to economies, we are perhaps inadvertently dooming elements of it to extinction. Only when ecologically sound management plans based on scientific studies are developed for resource extraction will the use of those resources be able to contribute to the conservation of biological diversity.

Conservation Areas

Early nature preserves were established in the tropics during colonial times primarily to serve the needs of big-game hunters or to protect watershed and timber resources. The colonial administrations created most exciting rain forest reserves by simply declaring government land to be a national park or by purchasing land from private owners, the same strategies followed to create national parks in North America and Europe.

The Mexican government owned more than 99 percent of the 528,000 hectares of rain forest, wetlands, and coral reefs that are now included in the Sian Ka'an biosphere Reserve on the Yucatan Peninsula, whereas the 100,000-hectare Guanacaste National Park in Costa Rica, an area of dry lowland tropical forest, was purchased largely from private landowners for $9 million contributed by a variety of conservation organisations, trusts, foundations, private donors, and government agencies.

The Monteverde Cloud Forest Reserve in Costa Rica, on the other hand, was created partially through a "debt-for-nature" swap-conservation organisations purchased part of the country's international debt and accepted conservation of rain forest acreage as payment. While these strategies have been effective in preserving land, they have been primarily focused on meeting national needs rather than the concerns of local peoples.

These traditional strategies create reserves that are essentially free from human disturbance. New strategies differ in that they emphasise the possibility of using the resource while protecting it from degradation. In the late 1980s, Brazil created a category of forest reserve known as the "extractive reserve," an area where local people can extract products on a small scale while still preserving a largely intact ecosystem. This form of biological reserve is closely associated with a social movement, begun in the state of Acre, which attempts to improve economic standards among Brazil's traditional peoples.

The first reserves were established for the extraction of rubber and Brazil nuts. Most of the rubber produced in the Amazon Basin is gathered in a way that does not destroy the trees, so the people who gather it are strongly opposed to any destruction of the rain forest. When ranchers who wished to clear the forest assassinated Chico Mendes, the movement's most visible leader and organiser of the local rubber tappers' union, there was such an outcry that the government eventually responded by creating the first major extractive reserves, which now total some 10 percent of the entire state of Acre. Over the last few years, however, the value of both wild-harvested rubber and Brazil nuts has fallen.

Ethnobotanists and taxonomists such as Douglas Daly of the New York Botanical Garden are working with local inhabitants of the reserves to identify other species that can be produced for regional and international commerce and provide income opportunities for the people who protect these areas of tropical forest. Individuals in north temperate countries are also helping to create extractive reserves. Such organisations as Conservation International and Cultural Survival have organised the marketing of forest products from reserves.

Maintaining an extractive reserve is not without its problems. The forest ecosystem may be damaged if economically important products are overharvested. An extractive reserve differs from a parcel of land that is simply public property in that a social structure is a key element of the reserve. Ideally, guidelines can be developed and rules and regulations enforced. While such rules do not exist or are strongly enforced elsewhere, it appears that most reserves are respected by the local people, particularly if they are established with an understanding of local culture and needs.

Sometimes the ecosystem of a forest reserve can be altered by protecting economically important species (or even augmenting them by strategic replanting) while other species-those whose value is not clear-are not protected with the same tenacity. Thus, some critics have argued, extractive reserves are able to protect only a portion of the biodiversity they contain. The lesson is that there is no one formula for protecting wildlands, especially in remote regions of the tropics. But in the global effort to assemble a jigsaw puzzle of conservation areas, ethnobotanical research can play an important role by helping to preserve and disseminate traditional knowledge. When this knowledge is applied, economic returns can accrue to those who make their living in the rain forest while still protecting it.

Indigenous peoples and Nature Preserves

Because indigenous peoples have seldom been involved in the planning process, conflicts have often arisen between them and Western-educated preserve managers. Conflict has been particularly acute over the management of Amboseli National Park in Kenya, which is overgrazed and thus capable of supporting far smaller animal populations than in the past. "The decline of Amboseli has little to do with the large number of tourists that used to visit it or with the increase in elephant population-excuses often used by the authorities, for which there is ample evidence to refute," writes David Lovett Smith, a former warden of Amboseli.

> The demise has been brought about, in my opinion, by inept management and a total lack of communication with the local people. . . .For it was the Masai people who themselves who looked after wildlife until governments and wildlife authorities took over its management, and, from the 1970s on, proceeded to mismanage it so badly.

Implicit in the authorities' explanation of Amboseli's degradation is the view that the Masai and their herds are inimical to efforts to conserve the ecosystem, rather than potential parts of the solution. But changing such a view requires reconsideration of some of the fundamental tenets of Western land ownership and management.

Consider the differences between Western and indigenous notions of property ownership among the Turkana, a pastoral African people living in the Rift Valley of north-western Kenya, about 20 kilometres north-west of Amboseli National Park. Like many indigenous peoples, the Turkana do not believe in private ownership of natural resources. Instead, the Acacia [Mimosaceae] trees that their goats feed on are administered as a communal trust. Villager elders ration feeding privileges, chasing away offending goats with sticks. In Western view, such communal trusts are inherently unstable. According to most Western theorists, such resources will inevitably be degraded, and the result will be what bioethicist Garrett Hardin has termed "the tragedy of the commons".

To forestall what they saw as the "inevitable" collapse of the Turkana grazing system, a team commissioned by the United Nations divided up the Turkana grazing areas into plots, which were then deeded to private individuals. The village elders' sticks were no longer required. Soon, however, all of the Acacia trees were all denuded. George Monbiot argues that while the indigenous system of communal ownership of the Acacia trees might not have been sustainable in a Western society, it worked for the Turkana.

A new and very important branch of ethnobotany might be termed "ethno-conservation biology"-the incorporation of indigenous conservation models into wildlands management. Attempts are now being made to document indigenous conservation strategies throughout the world.

Ethno-biomedical Forest Reserve in Belize

Belize is a small nation with a population of around 200,000. Vast tracts of forest still cover a significant portion of the country. A recent environmental profile of Belize estimated that more than 93 percent of the country could be classified as forest land, although this estimate was optimistic, for it excluded only urban areas and large-scale farming operations. In 1988 the Belize Ethnobotany Project was initiated to inventory, understand, and conserve as much ethnobotanical data as possible in a country that is undergoing rapid change, accompanied by loss of natural habitat and the erosion of existing cultures.

The project is a collaborative effort between The New York Botanical Garden's Institute of Economic Botany, and the Ix Chel Tropical Research foundation in the Cayo District with the Belize Centre for Environmental Studies, the Belize Zoo and Tropical Education Centre, the Belize College of Agriculture, and a host of other government and non-governmental organisations in Belize. This multitiered effort has linked the mutual interests and activities of local healers, farmers, students, ethnobotanists, and pharmaceutical researchers to the conservation of their main source of materials and ideas: the area's forests. These forests serve as both a classroom and a source of raw materials for local health practitioners.

An ongoing inventory of species and their uses focuses in the collection, documentation, and study of traditional medicines. Collection efforts in small villages and isolated forest regions have been linked to the Developmental Therapeutics Programme of the National Cancer Institute in the United States, supplying it with more than 2000 bulk samples for testing in its cancer- and AIDS-screening programme.

Supplementing the more familiar role of ethnobotany as a documentary science, the project seeks to renew interest in cultural knowledge and its transmission, particularly in the area of medical practices. It has focused on work with groups of elderly healers, most of whom have no apprentices and whose accumulated knowledge is in danger of being lost. The long-term interdisciplinary nature of the project has allowed in-depth work with traditional healers in efforts to understand disease concepts, healing traditions, and the uses of plants. This type of knowledge recovery has been described as "salvage ethnobotany".

The project helped local organisations convene four national traditional healers' meetings. The open forum provided by these meetings enabled healers from different cultural groups and geographic regions of Belize for the first time to exchange information about the medical uses of local and exotic plants. They discussed the importance of traditional healing, the central role of the healer as community health care provider, and the increasing difficulty of locating certain useful species.

In 1992 the Belize Association of Traditional Healers was formed and Rosita Arvigo of the Ix Chel Tropical Research Foundation was elected its president. Yet without plants, their work is impossible. As one of the healers, Hortense Robinson, said, "we can't do our work without the plants-it's like a mechanic without his tools. Just knowing what the name of the plant was won't help-you can't use the name to heal you." As part of the effort to conserve species that are important to the work of traditional healers, a 2400-hectare parcel of lowland tropical forest was given forestry reserve status in June 1993 at the suggestion of a government minister, Daniel Silva, who noted that Belize has a rich tradition of conservation reserves.

The reserve was intended to provide a source of medicinal plants as well as a place to teach apprentices. Funds for surveying and demarcating the reserve were provided by the Healing Forest Conservancy and the Rex Foundation. The forest, in the Yalbak region of Belize, contains a wide diversity of fauna as well as many useful medicinal plant species. As originally conceived, this "ethno-biomedical forest reserve" would serve as a site to promote ethnobotanical and ecological research in efforts to define harvesting regimes for sustainable extraction. Toward this end, a team of scientists is carrying out ecological inventories as well as experiments designed to learn at what rates bark and roots will regenerate after harvest.

Unfortunately, within a year after the reserve was established, the local government changed, and controversy arose over which group of local healers would be responsible for its operation. Various plans have been submitted to the Forestry Department and scientific experiments continue, but development of the educational and social component of the reserve's programme is currently on hold. Despite the best of intention, not all conservation efforts are immediately successful.

As habitat destruction and overharvesting are depleting the supply of medicinal plants in the forests of Belize, Rosita Arvigo and Gregory Shropshire of the Ix Chel Tropical Research Foundation have started a programme to develop horticultural nurseries in collaboration with Hugh O'Brien of the Belize College of Agriculture. As part of the programme, the subject of medicinal plants was introduced into the college curriculum.

The major goal of the joint project is to learn to propagate many of the commercially valuable plants currently harvested from the wild. The species differ wildly in their morphology and biology-some are easily reproduced weedy herbs while others are long-lived trees-so the task is complex. Local and regional businesses that depend on native plants will ultimately benefit from techniques being developed in the nurseries. The project also rescues plants threatened by development. A team from Ix Chel and the Belize College of Agriculture collects seedlings of rare or slow-growing trees from areas soon to be cleared for housing and transplants them to a "tree orphanage." Eventually they will settle the young trees in more secure areas, such as forest reserves and privately owned farmlands.

Ethnobotanical Conservation in Future

Many challenges face ethnobotanists in future years, particularly the rapid loss of biodiversity and the concomitant loss of indigenous knowledge systems. UCLA anthropologist Johannes Wilbert tells us that many years ago the Warao of Venezuela were highly amused when he carefully documented their traditional dances. When their grandchildren attempted the same dances three decades later, they turned to Wilbert to find out if they were doing them correctly.

Some contemporary critics fear that outsiders' studies of traditional knowledge is not without risk. Published reports of the use of a medicinal plant might create a demand for the resource, with riches flowing to all parties involved except the original owners of the knowledge. Others fear that if only the most sensational information is written down, the more mundane information-about food plants, for instance-may be lost. Clearly, one priority for the future is to involve indigenous colleagues in ethnobotanical research as coinvestigators and to train a new generation of people from a variety of cultures to initiate studies among their own people.

Increasingly, the local people involved in ethnobotanical research, especially the healers, are being credited as co-authors of scientific papers and receiving patent rights to discoveries that result from the information they have provided. In ethnobotanical research, anything less than equal treatment should be viewed as unacceptable.

Many Indigenous Peoples, Nation states and the Multinationals

Exploring tropical forests for new pharmaceutical leads is being promoted enthusiastically as an innovative way to save the forests. 'Bioprospectors' express optimism that they can help to implement the 1992 Convention on Biological Diversity by encouraging biodiversity conservation and stimulating capacity building in developing countries. Many indigenous peoples, though, are sceptical of existing bioprospecting agreements. Most agreements are based upon acceptance of two international trends which indigenous peoples feel threatened by. These are the standardisation of intellectual property rights law, and the extension of nation state sovereignty to encompass all biogenetic resources within national boundaries.

History of Bioprospecting

Bioprospecting is the collection and screening of biogenetic resources for industry. One of the first scientists to argue that such an activity could constitute the basis for a conservation strategy was Thomas Eisner of Cornell University, USA. In 1989 Eisner and his colleagues in the International Society of Chemical Ecology adopted the Göteborg Resolution, which stated:

> Natural products constitute a treasury of immense value to humankind. The current alarming rate of species extinction is rapidly depleting this treasury, with potentially disastrous consequences. The International Society of Chemical Ecology urges that conservation measures be mounted worldwide to stem the tide of species extinction, and that vastly increased biorational studies be undertaken aimed at discovering new chemicals of use to medicine, agriculture and industry. These exploratory efforts should be pursued by a partnership of developing and developed nations, in such fashion that the financial benefits flow in fair measure to all participants.

The recent resurgence of natural product-based research by the pharmaceutical industry is mainly due to technological advances. However, as Eisner acknowledges, bioprospecting is hardly new. Since 18th century 'gene hunters' from Europe and North America have 'discovered' botanical treasures in the tropics, and these have generated enormous wealth. Indeed, for several generations virtually everybody living in the biodiversity-poor North has been a beneficiary of free northward transfers of biological resources.

However, a number of influential conservationists believe that the world is undergoing an extinction 'spasm', and that many resources are vanishing before people become aware of their existence. This situation is attributed mainly to large-scale clearances of the biodiversity-rich tropical forests. Southern governments are frequently blamed for letting this happen, but respond that as long as Northern countries fail to share the benefits of commercial exploitation of biogenetic resources, conservation cannot be justified economically.

Intergovernmental negotiations, such as those resulting in the agreed text of the Convention on Biological Diversity, have to some extent accommodated the demands of multinationals seeking access to the resources of biodiversity-rich areas, and of Southern governments wishing to control what they regard as their biogenetic assets. Unfortunately, indigenous peoples have tended to be marginalised from these debates, as they are from by the types of partnership promoted by many bioprospecting enthusiasts.

Convention on Biological Diversity (CBD)

The Convention on Biological Diversity (CBD) came into force in 1993 and has been ratified by over 160 countries. There are three main objectives:

— the conservation of biological diversity,

— the sustainable use of its components,

— the fair and equitable sharing of the benefits arising out of the utilisation of genetic resources.

Agreeing a text acceptable to governments in the biodiversity-poor North and the multinational companies on one side, and the Southern governments on the other, proved to be a difficult and contentious process. The latter countries together possess most of the planet's biological wealth but require greater scientific, technological and financial capacities to exploit it.

The multinationals wanted continued free access to biological resources; Southern governments demanded technology transfers and benefit-sharing as conditions for access. To some extent Southern governments prevailed. Nevertheless, only the United States refused to sign at the Earth Summit, the other Northern countries deciding that it was an agreement they could live with, as did most multinationals (Note from the editor:

The USA have in the mean time signed the CBD Agreement, but it has not (yet) been ratified by their Congress). In many ways, then, the CBD is a framework agreement setting out the terms on which the world's biogenetic wealth is to be shared out between nation states in the South and industrial concerns mostly in the North, but with some potentially far-reaching concessions to indigenous peoples and local communities. Thus, while the sovereignty of nation states is now extended by treaty to all genetic resources within their borders, the intellectual property rights of corporations were recognised in the final version of the CBD, albeit in somewhat ambiguous language.

Genetic resources have been considered as part of the common heritage of humankind. Given the widely disparate abilities of countries to exploit these resources profitably, it was understandable that developing countries would wish to challenge this assumption. However, this extension of the domain of the nation state is highly problematic for indigenous peoples, because it appears to contradict international human rights law, according to which 'all peoples may, for their own ends, freely dispose of their natural wealth and resources'. Indigenous peoples fear that governments which enact legislation to implement the CBD will apply this principle unrestrictedly, and thereby violate their territorial integrity and resource rights.

Developing countries in recent years have been put under tremendous pressure to adopt intellectual property rights (IPR) regimes that accord with the standards which now prevail in North America, Europe and Japan. Discussions on the standardisation of IPR have taken place mainly during the Uruguay Round of the General Agreement on Tariffs and Trade (GATT), although the CBD is another important forum where IPR has been debated.

The Uruguay Round culminated in an international agreement commonly known as the 1994 GATT Final Act, which was signed in 1994 by 124 governments. It includes the Agreement on Trade-Related Aspects of Intellectual Property Rights (TRIPs). During negotiations the United States and several other Northern countries demanded protection for biotechnology products and processes in the broadest possible sense, including the patenting of life-forms.

As a result, although members may exclude from patentability 'plants and animals other than microorganisms, and essentially biological processes for the production of plants or animals other than non-biological and microbiological processes', TRIPs does not easily allow countries to prohibit the patenting of all life-forms, whether on moral or other grounds. Therefore, patenting of genetically-modified organisms and even human genetic material will probably be permitted by more and more countries. Indigenous peoples are outspoken in their opposition to such patents, considering them to be violations of the sanctity of life and contrary to their moral beliefs.

While compliance with GATT-TRIPs and the CBD requires states to implement domestic legislation consistent with their provisions, a potential conflict exists. Implementing GATT 1994 will tend to harmonise IPR law, while national legislation to implement the CBD in biodiversity-rich countries is likely to link access to biogenetic resources to expanded nation state rights, and perhaps indigenous rights, too. Two important and inter-linked issues arise here: the possibility of adapting IPR regimes so they can protect indigenous knowledge, and the question of intellectual and biological 'piracy'.

Conventionally, IPR are conferred upon individuals and corporate entities, but not upon communities. While intellectual property law can protect inventions of companies based upon the knowledge of local communities, the community knowledge itself cannot be protected. Furthermore, in the case of patents the geographic location of the source of the raw material used for the 'invention' has no bearing on the decision to accept or reject a patent application. These features of patents effectively support the act of 'biopiracy' in which indigenous peoples and local communities have their knowledge and resources taken and used without their authorisation by corporations and governments without their prior agreement.

One controversial instance of this is the case of the neem tree (*Azadirachta indica*), which has been used for centuries by Indian farming communities for numerous purposes, including protecting crops from insect pests. Two companies in the United States have patents for derivatives of the active principle without having compensated local farmers for their knowledge and prior use of the tree. It is estimated that the global market for neem-based bio-insecticides will reach US$ 50 million per annum by the year 2000.

Indigenous Perspectives

Given the sources of knowledge in such cases as neem, the fairness of IPR law is being questioned by indigenous peoples. The Draft UN Declaration on the Rights of Indigenous Peoples expresses the concerns, demands and aspirations of hundreds of indigenous peoples' organisation around the world.

Evidently, indigenous peoples interpret their cultural and intellectual property broadly, so that these encompass much more than knowledge, but also their cultural heritage, their biological resources, and even their cells and DNA.

In fact, indigenous peoples have their own regimes to regulate access to and control over knowledge and resources that are often more sophisticated than those based on IPR or national sovereignty. According to the North American indigenous peoples' organisation, the Four Directions Council:

> Indigenous peoples possess their own locally-specific systems of jurisprudence with respect to the classification of different types of knowledge, proper procedures for acquiring and sharing knowledge, and the rights and responsibilities which attach to possessing knowledge, all of which are embedded uniquely in each culture and its language.

For this reason, the Four Directions Council argues that:

> Any attempt to devise uniform guidelines for the recognition and protection of indigenous peoples' knowledge runs the risk of collapsing this rich jurisprudential diversity into a single 'model' that will not fit the values, conceptions or laws of any indigenous society. A better approach ... would be for the international

community to agree that traditional knowledge must be acquired and used in conformity with the customary laws of the peoples concerned.

For indigenous peoples, then, protection of knowledge and resources, and continuation of customary law and practice, are central to maintenance of their cultural identity. Therefore, control over these is an aspect of human rights. This needs to be understood by all governments, companies and other institutions before they enter into negotiations for the use of biogenetic resources on the territories of indigenous peoples. These negotiations should certainly involve indigenous peoples. Often they do not. The following case, despite a generally favourable image, has attracted criticism for failing to accommodate local needs and concerns.

National Biodiversity Institute (INBIO)

The National Biodiversity Institute of Costa Rica is frequently hailed as a model bioprospecting institution, and its agreements with pharmaceutical companies are regarded as examples for other countries to follow. It is true that INBio's activities and collaborations have some positive features, including capacity building and benefit-sharing. In other respects, INBio and its agreements are inappropriate for other countries and may be inappropriate for Costa Rica, too.

INBio is a private non-profit organisation which was established in 1989 with the blessing of the government. Its main objectives are to carry out a species inventory of the country -- which is estimated to have 4% of the world's biological diversity -- and to explore the commercial potential of the country's biological resources.

Costa Rica's 1992 *Conservation of Wildlife Law* declares national sovereignty over the biological diversity of the country. This is not equivalent to nationalisation of all biogenetic resources in the country. Nevertheless, the State has the exclusive right to grant permits to investigate, collect and exploit the country's biological diversity in the publicly-owned conservation areas, which make up 25% of the whole country. The Ministry of Natural Resources, Energy and Mines (MIRENEM) is authorised to grant bioprospecting permits in these areas, and INBio has such a permit.

INBio's first and best known agreement with a multinational company was the one with Merck, which began in 1991 and has been renewed up to the present. Merck gave INBio an advance payment of US$1 million, equipment for an extraction laboratory, and a guarantee of royalties in case a product is derived from any of the extracts (of plant and insect specimens) which INBio will transfer to Merck. 50% of the royalties will go to the government's National Parks Fund through MIRENEM.

The agreement has certain positive aspects. First, the advance payment, which is not a common feature of such agreements; second, INBio's freedom to supply extracts to other companies, including those given to Merck if two years have elapsed since Merck received them; third, INBio has control over taxonomic information on all its samples. Thus, if an extract shows promise for Merck to continue its research it will have to approach INBio again to obtain more material.

However, before promoting the Costa Rican approach as a model, certain points merit consideration:

— The government has granted INBio prospecting rights to explore lands and biogenetic resources over which the State has jurisdiction. Although Costa Rica is more culturally homogeneous than its neighbours indigenous peoples exist whose inalienable rights to lands and resources pre-date the existence of the country. Therefore if indigenous peoples are not parties to negotiations, an argument can be made that both the government and INBio are effectively usurping the territorial and resource rights of indigenous peoples.

— The government and INBio are the beneficiaries of the Merck agreement. The only way that local communities appear to gain in any direct sense is through the training of a small number of local 'parataxonomists'. Furthermore, INBio will not contribute at all to revitalising local knowledge traditions because it professes to have no interest at all in such knowledge. Nor, one can only conclude, in the cultural diversity that underpins such traditions. According to Maurice Iwu, a prominent Nigerian scientist:

> The kind of deal that was done in Costa Rica cannot be done in Nigeria... The kind of deal that India and Nigeria [etc.] will negotiate has to recognise the multi-ethnic nature of our own societies.

— A company like Merck will tend to seek cures for therapeutic groups that are major concerns in developed countries, but which may be relatively less important in developing countries where other diseases and infections take higher priority. According to Iwu:

> Screening strategies should include parasitic infections and diseases (e.g., malaria) to aid in the search for new treatments for diseases of primary concern to source country inhabitants, and proper objectives should not be limited to the generation of pure chemical isolates as pharmaceutical leads, but should include the standardisation of phytomedicines for the benefit of traditional healers and their patients. The inclusion of such measures will contribute to an improvement of the quality of life of source country inhabitants.

Self-determination of Bioprospecting Rules

Indigenous peoples are concerned that bioprospecting expeditions, even those carried out with good intentions, will infringe their rights. To a large extent, the rules of the game are those of multinationals and nation states, and indigenous peoples still find it difficult to get their voices heard. In response, a growing number of groups, such as the Kuna of Panama, the Awa of Ecuador and the Inuit of the Arctic have developed their own regulations which visiting scientists must adhere to.

In India local communities are setting up community registers of local knowledge of biodiversity to revitalise traditional knowledge systems for the benefit of future generations and to protect them from piracy. Such initiatives are being carried out independently of governments and companies. Other indigenous peoples have declared their opposition to bioprospecting. Participants at the UN Development Programme-sponsored Regional Consultation on Indigenous Peoples' Knowledge and Intellectual Property Rights held in April 1995, called for a moratorium on bioprospecting in the Pacific region and urged indigenous peoples:

Indigenous peoples demand that their right to self-determination be recognised *before* they are prepared to enter into negotiations over access to their territories and resources. Given the vital role indigenous peoples play in the conservation of biodiversity it is vital for all that national laws to implement the CBD and GATT be enacted which uphold this right.

8

Industrial Utilisation of Medicinal Plants

Most developing countries are endowed with vast resources of medicinal and aromatic plants. These plants have been used over the millennia for human welfare in the promotion of health and as drugs and fragrance materials. This close relationship between man and his environment continues even today as a large proportion of people in developing countries still live in rural areas. Furthermore, these people are precluded from the luxury of access to modern therapy, mainly for economic reasons.

About 80% of the population of many developing countries still use traditional medicines for their health care. Modern pharmacopoeia still contain at least 25% drugs derived from plants and many others which are synthetic analogues built on prototype compounds isolated from plants. China, India, Sri Lanka and a few other countries have officially recognised the use of traditional medicines in their health care delivery systems.

A major factor impeding the development of the medicinal plant based industries in developing countries has been the lack of information on the social and economic benefits that could be derived from the industrial utilisation of medicinal plants. Except for the use of these plants for local health care needs, not much information has been available on their market potential and trading possibilities. As a result, the real potential of these plants has not been exploited by the governments or entrepreneurs.

The demands of the majority of the people in developing countries for medicinal plants have been met by indiscriminate harvesting of spontaneous flora including those in forests. As a result many plant species have become extinct and some are endangered. It is therefore necessary that systematic cultivation of medicinal plants be introduced in order to conserve biodiversity and protect threatened species.

Systematic cultivation of these plants could only be initiated if there is a continuous demand for the raw materials. It is therefore necessary to establish processing facilities preferably in the vicinity of cultivations in order to create a demand and assure farmers of the sale of raw material. Thus cultivation and processing should be started simultaneously. In cases where parts of mature plants from forests are to be harvested, sustainable harvesting techniques are needed to protect these plants.

The promotion and development of processing of plant-based products have been given a fresh impetus due to certain ground realities:

— Green consumerism and the current resurgence of interest in the use of "Naturals" in developed countries.

— Free market economy bringing in more openness and expanding markets and demand for new resources, materials and products.

— A growing acceptability of the social responsibility of minimising socio-economic inequalities in favour of rural people resulting in creating additional job and income opportunities for poor people.

— Poor economic conditions in developing countries restricting import, thereby placing increased reliance on medicines using local plant resources.

— Increasing awareness regarding biodiversity conservation and the sustainable and protective use of plant resources.

— Search for new phytopharmaceuticals for the prevention and cure of deadly diseases such as cancer and AIDS.

In 1981, the United Nations Industrial Development Organisation (UNIDO) established a programme for the systematic utilisation of this renewable natural resource for the benefit of the populace of developing countries. The programme aims for initiating development projects for increasing the industrial output of developing countries in the field of locally-used herbal drugs and phytopharmaceuticals which affect the economies of both developing and developed countries.

Some of the constraints associated with the processing of medicinal plants which may result in reducing their competitiveness in global markets and which have to be remedied are:

— Poor agricultural practices,

— Poor harvesting (indiscriminate) and post-harvest treatment practices,

— Lack of research on development of high-yielding varieties, domestication etc.

— Poor propagation methods,

— Inefficient processing techniques leading to low yields and poor quality products,

— Poor quality control procedures,

— High energy losses during processing,

— Lack of current good manufacturing practices,

— Lack of R & D on product and process development,

— Difficulties in marketing,

— Lack of local market for primary processed products,

— Lack of trained personnel and equipment,

— Lack of facilities to fabricate equipment locally,

— Lack of access to latest technological and market information.

Systematic cultivation of many medicinal plants needs specific cultural practices and agronomical requirements. These are species-specific and are not only dependent on soil, water and climatic conditions. Hence research and development work has to be done to formulate Good Agricultural Practices which will include proper selection and identification, propagation methods, cultivation techniques, harvesting, step-wise quality control of raw material up to processing stage, post-harvest treatment, storage and safety. These aspects have to be incorporated into protocols for the cultivation of medicinal plants.

Organic farming is another practice that is gaining wide acceptance as world demand particularly in developed countries for organically grown crops is rapidly on the increase. Farmers have to be trained in all aspects of organic farming of medicinal plants and herbs including obtaining certification from associations that do the monitoring starting from cultivation to final harvesting. Organic farming which is labour-intensive gives the developing countries the comparative advantage to be competitive.

Traditional and Modern Technology

The medicines for internal use prepared in the traditional manner involve simple methods such as hot- or cold-water extraction, expression of juice after crushing, powdering of dried material, formulation of powder into pastes via such a vehicle as water, oil or honey, and even fermentation after adding a sugar source.

Traditional herbal medicines were produced using age old methods by the practitioner him/herself who was able to identify the correct plant species. This practice of the traditional practitioner dispensing his own medicines is being gradually shifted to herbal drug stores which are profit-oriented. As a result, there is no guarantee of the authenticity and quantity of plant material used in the preparations.

The quality of traditional medicines so produced vary widely and may not even be effective. Therefore, there is a need to select proper and appropriate technologies for the industrial production of traditional medicines such that the effectiveness of the preparation is maintained. Traditional methods used have many disadvantages which could be corrected by selecting the suitable technologies. It has to be stated that the traditional methods were dependent on the status of technology that was available at that time. It therefore follows that these can be modified and improved using the technologies available today to make them more effective, stable, reproducible, controlled and in dosage forms that can easily be transported or taken to office.

Hence the introduction of appropriate, simple and low-cost technologies should be encouraged maintaining as much as possible the labour-intensive nature of such activities, conservation of biodiversity through small-scale production and preservation of cultural knowledge. Use of sophisticated modern technology will alienate the traditional practitioner as he has no control over such production methods. Even in the use of appropriate technologies, the practitioner who produces these drugs has to be educated about the advantages of using such production and quality control methods.

One major concern in introducing modern technology for the production of traditional medicines is whether the final preparation will be acceptable to the practitioner who has sole faith in extemporaneous preparations. This problem has to be overcome by a process of education, whereby the disadvantages of the old methods and the advantages of the new methods can be imparted.

The value of medicinal plants as a source of foreign exchange for developing countries depends on the use of those plants as raw materials in the pharmaceutical industry. These raw materials are used to:

- Isolate pure active compounds for formulation into drugs (quinine, reserpine,digoxin,etc.)
- Isolate intermediates for the production of semi-synthetic drugs
- Prepare standardised galenicals (extracts, powders, tinctures, etc.)

If one is to produce known pure phytopharmaceuticals used in modern medicine, more processing stages and more sophisticated machinery are required. Furthermore, safety and pollution aspects have to be considered. Most of these processes and formulations are patent-protected. Even transferring technology through contractual agreements and payment will not be of much help unless there is a large local demand for these drugs. Often the drugs so produced are more expensive than world market prices owing to the limitations of the economy of scale of production.

Certain plants are rich sources of intermediates used in the production of drugs. The primary processing of parts of plants containing the intermediates could be carried out in the country of origin thus retaining some value of the resource material. For example, diosgenin (from *Dioscorea* sp.) and hecogenin (from sisal) used in the production of steroids can be commercially produced in the countries of origin where there are steady supplies of sufficient raw materials.

Processed products (galenicals) from plants could be standardised fluid/solid extracts or powders or tinctures. Standardised extracts of many plants (e.g. *Aloe* species, *Atropa belladonna, Cassia angustifolia, Capsicum annuum, Centella asiatica, Cephaelis ipecacuanha, Digitalis species, Commiphora mukul, Panax ginseng*) are widely used in health care. Some of these have to be formulated for incorporation into modern dosage forms. New formulations require some development work, particularly on account of the nature of the processed products. Plant extracts are difficult to granulate, sensitive to moisture and prone to microbial contamination. Hence the types of excipients to be used and the processing parameters have to be determined.

Most of the activities of UNIDO in the industrial utilisation of medicinal plants have involved: the production of standardised traditional medicines, galenicals and extracts, the formulation and development of dosage forms, the development of new preparations based on the traditional pharmacopoeia, research and development in processing and formulation and basic chemical and pharmacological studies. Processing using clean and pollution free technologies have been introduced.

Development of process parameters has to be carried out at a pilot scale in order to be upscaled to industrial production. Many R&D institutions and universities in developing countries do not have such facilities and are therefore unable to pass on their R&D findings to the industry. UNIDO has played a significant role in achieving university-industry interaction by designing a polyvalent pilot plant which could be used for many unit operations needed for the production of plant-based products.

Detailed engineering drawings with specifications and bills of quantities are published in a separate UNIDO publication. The pilot plant has enabled the R & D work to be demonstrated to produce final products which could be market tested. Universities in turn have been able to realise the hidden potential of industrialisation of their R & D work. Any institution which wants to initiate downstream processing of medicinal plants and other non-wood forest products should possess facilities for pilot scale production.

Improved Technologies

Improved methods for the processing of medicinal and aromatic plants and new techniques for quality assessment are being developed rapidly and continuously and they have to be introduced to developing countries if they are to forge ahead to keep up with recent developments and new international requirements. One such method of recent development is Supercritical Fluid Extraction of natural products as an alternative for solvent extraction. It is now used both for processing of phytopharmaceuticals and other plant products and for the removal of trace amounts of contaminant substances such as pesticides, toxins and surfactants.

Quality of the Raw Materials

The control of the quality of the raw materials, finished products and of processes is an absolute necessity, if one is to produce goods for world markets and human consumption.

International Standard Specifications exist for some processed products and some countries and buyers have their own requirements. The quality requirements for medicinal plant preparations are stringent in terms of content of active principles and toxic materials. Whereas the production of traditional medicines for local use does not require such stringent standards, what is produced will be a much more improved version of the already produced medicines using traditional methods.

Quality has to be built into the whole process beginning from the selection of propagation material to the final product reaching the consumer. It is therefore a management system where all steps involved in the industrial utilisation process have to be properly and strictly controlled to produce the desired quality products. All elements of Total Quality Management(TQM) have to be introduced in any industrial project. The requirements for ISO 9000 certification and Good Manufacturing Practices(GMP) have to be introduced and the personnel trained so that enterprises could introduce the proper systems needed for certification. Furthermore eco-audit procedures (ISO 14000) leading to eco-labelling will be required for safeguarding environmental damage.

Marketing in Developing Countries

Marketing is an unsurmountable problem besetting the development of the plant-based industry in developing countries and marketability will be a crucial factor in determining the failure or success of these industries. The market outlets can be for local use and for export. As for local use, some products could reach the consumer directly while others have to be either further processed or used as secondary components in other industrial products. Hence user industries have to be promoted so that locally produced extracts can be used to save foreign exchange needed for importation of such additives.

Further processing to yield value added products will be limited by the local demand situation unless they could be produced at prices to be competitive in the world market. Even if the cost of production is low and quality of the products is good, substantial market promotion has to be undertaken in order to penetrate the world market.

Market tie-ups with progressive entrepreneurs from the developed world would be a convenient and a realistic option for securing markets for the finished products. Joint ventures or trade agreements could be initiated with companies that are expanding their production to meet the ever increasing demand for green products.

Parallel Research and Development

Industrial development requires parallel research and development. Research in chemistry and bioactive components of indigenous flora of developing countries has been ongoing for quite some time, funded by multi-/ bi- lateral aid or non-governmental donor organisations. A systematic and a concerted approach to this activity has not been maintained for want of sophisticated equipment and high-cost chemicals. Much of the research has been mainly academic. The concept of applied research in the industrial use of plants has not received much attention.

Research in support of industrial development encompasses all activities ranging from the development of superior propagation materials, agrotechnology, low cost and efficient processing technologies to improve quality and yield, new formulations to new products and the marketing of finished products.

Property Rights

Many developing countries do not have procedures to register medicinal plant preparations although they are widely used for the health care needs of the majority of people. The regulations if any, are very stringent requiring the same standards expected of modern medicines. WHO has recently published guidelines for the assessment of herbal medicines taking into account the long and extensive usage of them. These guidelines should encourage developing countries to relax some of the current regulations to be realistic in recognising the role of traditional medicines in the health care delivery of their countries.

The vital question of property rights to developing countries for the use of know-how and genetic resources in the development of modern drugs in developed countries has been discussed in many forums but without a final solution.

Technical Information

As the use of plant-based drugs has declined with the introduction of synthetic drugs especially in developing countries, the need to collate all available knowledge on medicinal plants and their uses has become imperative particularly because of the recent revival of interest in the use of "natural" medicines. In view of the greater demand for information on traditional uses and proper identification, and the enormous volume of information being generated, a user-friendly information storage and retrieval system will be of considerable importance. As such a computer database would be of invaluable assistance to developing countries in R & D work on industrial utilisation of local plant resources.

Data on the medicinal plants are also available in international journals and a number of databases, the premier being Napralert established at the University of Illinois in Chicago with WHO assistance. Many developing countries lack the resources to subscribe to research journals or acquire access to these databases. In fact, the data required by the scientific personnel in developing countries with respect to technologies and methods used for processing and formulation of medicinal plants are not readily available in the literature nor in the databases as some of these are patent-protected.

9

Conservation of Medicinal Plants

Today many medicinal plants face extinction or severe genetic loss, but detailed information is lacking. For most of the endangered medicinal plant species no conservation action has been taken. For example, there is very little material of them in genebanks. Also, too much emphasis has been put on the potential for discovering new wonder drugs, and too little on the many problems involved in the use of traditional medicines by local populations.

For most countries, there is not even a complete inventory of medicinal plants. Much of the knowledge on their use is held by traditional societies, whose very existence is now under threat. Little of this information has been recorded in a systematic manner. Besides the identification and selection of medicinal plants for use in health services, there is the potential that plants hold as an inexhaustible reservoir for the identification and isolation of useful chemical compounds for syndromes such as AIDS, for which there is yet no known cure.

In the light of this situation, WHO, IUCN, and WWF decided that it would be timely to collaborate in convening an International Consultation on the conservation of medicinal plants, bringing together leading experts in different fields to exchange views on the problems, determine priorities and make recommendations for action. The experts at the meeting included administrators and policy-makers in health and conservation, and covered the disciplines of ethnomedicine, botany, education, pharmacology, nature conservation and economics. For IUCN and WWF, this meeting was an important part of their Plant Conservation Programmes.

The consultation took place in Chiang Mai, Thailand, on 21-27 March 1988, with the Ministry of Public Health of the Royal Thai Government as host. A wide range of topics was covered, which included a review of medicinal plant policies (utilisation and conservation) in individual countries; the need for information systems, including databases; and the part that botanic gardens can play in the cultivation and conservation of endangered medicinal species.

A lively and stimulating exchange of views took place between conservationists, scientists and health administrators, who were meeting for the first time in the same forum. The participants prepared and issued "The Chiang Mai Declaration— Saving Lives by Saving Plants"- which affirms the importance of medicinal plants and calls on the United Nations, its agencies and Member States, as well as other international organisations, to take action for the conservation of medicinal plants.

The meeting divided for part of its time into several working groups to develop a set of guidelines, primarily for governments, outlining in concise form what needed to be done. These guidelines are presented here and will be disseminated widely to governments and relevant institutions throughout the world for adaptation to local situations. Professor Vernon Heywood, formerly of IUCN, prepared the outline, which was developed by the participants. Hugh Synge coordinated the completion of the guidelines after the Conference taking into consideration inputs from participants, and prepared the final text with Olayiwola Akerele, formerly of WHO, and Vernon Heywood.

The Forty-first World Health Assembly (1988) in its resolution WHA41.19 drew attention to the Chiang Mai Declaration and endorsed the call for international cooperation and coordination to establish a basis for the conservation of medicinal plants, so as to ensure that adequate quantities are available for guture generations. This places medicinal plants, their rational and sustainable use, and their conservation, firmly in the arena of public health policy and concern.

The Chiang Mai meeting covered only plants, but traditional practitioners also use animals and minerals in their remedies. Indeed, the threat to animals by medicinal uses can be even greater than the threat to plants. For example, the use of rhino horn in traditional Chinese medicine has decimated the population of Asian rhino over the last two decades. Any future activity on conservation and medicine should cover not only plants but also animals and minerals.

No single sector, private or public, can undertake the conservation of medicinal plants alone. the job requires a team effort, involving a wide range of disciplines and institutions.

The best was to start and orchestrate such a process is for each country to prepare a national strategy for the conservation and sustainable use of its medicinal plants. The process of preparing a Strategy will help in:

— developing a consensus on what needs to be done;

— assigning tasks to different institutions;

— motivating participants to undertake the tasks; and

— monitoring progress.

In such a process, the Guidelines presented here would serve as a checklist of tasks to be done. Those preparing a national strategy should consider all the tasks in the Guidelines, but will undoubtedly wish to give more weight to some tasks than others, depending on local needs and circumstances.

The Guidelines assign each task to a target group (e.g., the task of ex situ conservation is proposed to botanic gardens). However, the assignation of tasks to specific government departments, government agencies and non-governmental bodies may vary greatly from one country to another. Preparing a national strategy is the way in which countries can ensure that each task is assigned to the institution best able to carry it out.

One way of starting the process would be to hold a regional and/or national workshop, bringing together experts on different aspects of the subject to asses the situation, define objectives, set priorities, and draw up a plan of action. In developing and implementing the strategy, it is essential to work in partnership with those who use medicinal plants— herbalists, plan collectors, health workers and local people, for example. NGOs should also be brought in from the beginning.

An advantage of a national strategy is that it makes it easier for the donor community to identify activities that would match their funding priorities. Already WHO collaborates with and supports Ministries of Health in the development of programmes on utilisation of medicinal plants, IUCN helps many countries prepare National Conservation Strategies, which include the conservation of plants, and WWF funds many projects to conserve plants.

A national strategy is also a good way to secure the involvement and continued participation of the different disciplines involved. One of the remarkable features of the Chiang Mai meeting was the synergy and complementarily of efforts that emerged from the various disciplines present. For example, conservationists learned about how health workers actually used medicinal plants; health policy-makers learned about the efforts of conservationists in maintaining areas of natural vegetation the ultimate source of their medicinal plants. This synergy was not only very stimulating for the participants, but it also proved very productive and rewarding for the results of the meeting.

At the Chiang Mai meeting, it was agreed that the main role of international organisations should be to encourage and support the work of individual countries, in particular in designing and implementing their national programmes on the conservation of medicinal plants. International organisations can also help by developing guidelines and monitoring their implementation. In particular, they can promote the exchange of information and expertise and the transfer of technology within and between countries, and so help countries to learn from each other's experience.

Since the meeting, WHO has continued to expand its work on medicinal plant utilisation, through promotional activities and development of specific guidelines. IUCN has continued to develop its policies for conservation of medicinal plants, both in situ and ex situ. WWF has continued its joint project with the Ministry of Indigenous Medicine in Sri Lanka as a practical model of medicinal plant conservation, commissioned a report by Conference participant Dr Tony Cunningham on the threats facing medicinal plants in Africa, and funded a range of ethnobotanical studies, mainly in rainforest countries. Also since the meeting, Botanic Gardens Conservation International, set up by IUCN and now an independent organisation, has been encouraging botanic gardens to conserve medicinal plants, particularly *ex situ,* a policy supported by recommendations of a number of meetings hosted by WHO.

Participants felt that the range of international organisations involved in the conservation of medicinal plants should be broadened. UNESCO, for example, could contribute through its Man and the Biosphere programme, especially through utilising the extensive network of biosphere reserves for conservation of medicinal plants. FAO has already been active on this issue and is one of the three sponsors (with the International Society for Horticultural Science— SIHS— and the Research Institute for Medicinal Plants) of a Newsletter of Medicinal and Aromatic Plants. The UN Industrial Development Organisation (UNIDO) also has activities on medicinal plants, in particular on transfer of technology for their genetic improvement.

One task that should be done at the international level is the development of a common design for databases on the conservation and sustainable use of medicinal plants. This should involve lead agencies such as WHO, FAO, UNIDO, UNESCO and IUCN. Within the limits of their financial and other resources, WHO, IUCN and WWF will consider requests from countries for assistance to implement these Guidelines.

Basic Studies

Ethnobotany

The natural vegetation of the world is disappearing or being altered at an alarming rate. Many cultures that have lived close to nature, depending on its products for **their** needs, are suffering rapid cultural, social and economic change. These people had a deep understanding of the properties of their local plants— a knowledge that is itself endangered.

Ethnobotany, the study of how people in traditional societies use plants, has great potential to provide new and useful plant products for the benefit of the world. Many of the plant extracts used in western medicine were discovered through their uses in traditional societies, though not necessarily for the same purpose. Equally important, however, is how ethnobotany can help local communities adapt to changing circumstances.

The practices of ethnobotany are themselves being modified to ensure that the rights of traditional peoples to their knowledge are safeguarded and that these people benefit from any commercial discoveries made from their knowledge.

The great knowledge on plants of traditional people in predominantly wilderness areas, such as of the Yanomani in Brazil, has received much public attention, but the Chiang Mai meeting also drew attention to the herbal knowledge of the urban poor in cities across the developing world. These people brought from their villages to the cities much valuable knowledge on herbal remedies that is rarely studied. Indeed, modern health care can benefit not only from the remarkable knowledge of indigenous peoples but also from the traditional practices found in virtually all cultures around the world.

For many years, WHO has been encouraging countries to re-examine their systems of traditional medicine and practices. Such an examination is much easier in systems of medicine for which the philosophy and educational content are well documented, such as Ayurveda or Unani, than in systems of traditional medicine handed down from one generation to another by word of mouth. In spite of the difficulties, the examination of the latter systems holds great promise for the future benefit of mankind.

Each country should identify and support one or more institutions to plan, coordinate and implement ethnobotanical surveys.

So far, most ethnobotanical surveys have been carried out by individuals, rather than by institutions. If the useful information of traditional peoples is to be documented before it is too late, ethnobotanical activities must be broadened and accelerated. To do this, the primary responsibility should move from the individual researchers to selected institutions, who can then provide the support, encouragement, coordination and implementation that are needed.

The selected institution(s) should implement a nationwide programme of surveys on the use of plants for medicinal purposes in traditional societies.

The teams carrying out the surveys should be multi-disciplinary, and should include the traditional practitioners as part of the team, and not treat them as people to be interviewed. To carry out this guideline, in the case of virtually all countries, training in ethnobotany will be needed to provide the necessary personnel. It is worth noting that the guideline only covers the use of plants as medicines. Although the term "ethnobotany" is often used in this context, the correct term is "ethnopharmacology", which is the subset of ethnobotany that covers the medicinal uses of plants.

The data on ethnobotany should be catalogued and analysed but only disseminated in such a way that the communities providing the data would receive benefits from any commercial use of the information.

The data should be catalogued, preferably using computer databases, to permit comparisons from one study to another, and to ease retrieval of the information. The communities that provide the information should receive benefits from its commercial exploitation. The ownership of genetic resources— and of information derived from them— is a controversial issue and, since it was not discussed at the Chiang Mai Consultation, no specific recommendations are made on it.

Some people regard genetic resources as an important part of a nation's heritage that should only be made available in return for financial compensation, while others believe there should be no restrictions on access and that genetic resources should be freely traded. A number of countries have requested WHO and related UN agencies to prepare guidelines on this issue, which is also addressed by the Biodiversity Convention signed at the UN Conference on Environment and Development.

It is particularly important to exchange information both within and between countries. National databases would facilitate this, though data that could be of commercial value could be "hidden" until appropriate patenting was complete. At the international level, the information may be disseminated through the expanding network of the WHO Collaborating Centres for Traditional Medicine.

The ministry of health should incorporate proven traditional remedies into national programmes of primary health care.

This guideline is in accordance with a policy recommendation made at Alma-Ata that called for, inter alia, the accomodation of proven traditional remedies in national drug policies and regulatory measures.

The Ministry of Health should be responsible for approving selected plant remedies for use in the health service, as well as banning the use of dangerous plants and products.

Traditional Health Practitioners (THPs) should constitute themselves into National Bodies

Such national bodies are crucial to the process of introducing traditional practices into national health care. They would facilitate information from the Ministry of Health to reach THPs, and the experiences and problems both of individual and groups of THPs to reach the Ministry. In particular, such a body could provide information from which an "early warning system" could be devised to show which medicinal plants are becoming scarce and so in need of conservation action.

Medicinal Plant Identification and their Abundance.

To use and conserve medicinal plants effectively, it is vital to know precisely which are the species concerned, what are their correct names, and where they grow. There is no accepted global list of the over 20,000 medicinal plants in use today. In many cases medicinal plants have been mis-identified. For these and other reasons, any country's programme to use and conserve medicinal plants should include a stock-taking to identify its medicinal plants, outline their distributions and assess their scarcity or abundance.

Each Country should have

- an adequate national herbarium with a botanical library, so that plant material can be adequately identified and stored within the country; and
- a cadre of well-trained botanists able to identify plants, who can staff the herbarium and other botanical institutes and departments.

Botanists identify plants using herbarium specimens, that is dried, pressed plants in folders or mounted on paper or card. The advantage of herbarium specimens is that they take up little space and that they last for hundreds of years. Using them, a botanist can compare a specimen collected with hundreds of other specimens whose identities have been verified by specialists over many years. Therefore, samples of medicinal plants should be collected and prepared as herbarium specimens. The specimens should include all the organs required for accurate naming, such as flowers, fruits, seeds, roots and normal leaves.

Specimens should be adequately labelled. The label should include a unique reference consisting of the collector's name and a sequential number, e.g. Smith 2568. Material from the same collection, such as microscope slides, should always carry the same reference. The label should also include full details of the locality and habitat, as well as those characters of the plant which cannot be seen in the dried specimens, e.g. the height of trees, the colour of flowers.

The specimen should be identified with the scientific (Latin) name of the plant. In practice, there is a high level of mis-identification, both of herbarium specimens and even in the literature. Therefore the determination should be checked by a competent taxonomist; this requires access to a major herbarium and to reference literature. It is estimated that there are around 250,000 species of flowering plants in the world, and many of them need specialists for their identification. Groups of plants for which specialists are most often needed include orchids, palms and grasses.

One problem in naming plants is the occurrence of synonyms— more than one scientific name for the same plant. The correct name is the one that conforms with the International Code for Botanical Nomenclature, but in some cases botanists may disagree over which name is correct, in others well-known names may be overturned for technical reasons and in others botanists may not know which are the correct names without laborious nomenclatural research.

More frequently, however, botanists may disagree over the taxonomic position or rank of a plant, so leading to more than one name for it, each equally valid. To ensure consistency of names, especially when plants are mentioned in legal texts, it is recommended that the national herbarium (or equivalent organisation) should prepare a list of Standard Names, which would then be followed in legislation and in the literature.

There are even cases, albeit rare, where the same scientific name is used by two different authors for two quite different plants. For this reason, the name of the botanist or botanists (called the "authority (ies)") who coined the name should be added after it (except when a list of a Standard Names is followed).

Vernacular names are useful and important, but are not always easy to correlate with scientific names. One plant species can have many vernacular names, each used in a different part of the plant's range. Conversely, one vernacular name may refer to different plant species in different areas. Nevertheless, vernacular names are often a useful clue to the scientific identity of a plant, and are needed in communicating back to the practitioners who use the plants.

Although the species is the normal unit of plant classification, botanists may further divide species into subspecies (major subdivisions, usually on a geographical basis), varieties (small, local or otherwise distinctive variants) and forms (usually individual genetic variants within populations). These subdivisions are particularly important for medicinal plants, as the medical effect of the plant may vary from one variant to another. Indeed, many species consist of a range of distinct chemical forms, types or races, which cannot be recognised one from another by their visible characters, but have markedly different chemical properties, which may influence their pharmacological content.

Botanical Institutions should make a Catalogue of all the Plant Species used for Medicinal purposes in the Country.

The following data should be included on each species as far as practicable:

- Its scientific (Latin) name, and its vernacular name(s);
- Its geographical distribution;
- Its scarcity or abundance, population size and conservation status;
- Its habitat;
- Description of the part of the plant used (e.g. stem or leaf) and how collected;
- Its use(s) in medicine, traditional or otherwise;
- Its occurrence in protected areas;
- Whether cultivated or not, and the availability and location of ex situ germplasm.

Where a modern comprehensive Flora of a country or region exists, this should be followed to ensure ease of reference, particularly for legislators. Line drawings or photographs should be included wherever possible.

Most developed countries have prepared lists of their threatened plants, often in the form of "Red Data Books".

However, most developing countries, especially those with rich tropical floras, have not been able to do so; this is due to a lack of basic field information on the status of the individual plant species, itself caused in part by the very richness of the floras, the lack of trained botanists, and often the speed at which natural vegetation is being lost.

However, in the case where a country cannot yet produce a lis opf all its threatened plants, it should be able to determine which medicinal plants are threatened, simply because these plants will be better known in the field and in commerce than other species. In determining plants as threatened, the criteria of the IUCN categories of threat should be followed. These rank the species in large subjective categories that measure the degree of threat to the species, and hence the likelihood of its local or complete extinction.

Putting the information into computerised databases has the advantage of enabling retrieval in many different ways yet permitting a constant process of refining and updating. The use of international data standards and agreed transfer formats enables institutes to share and contribute information more easily with each other, and greatly saves in the time and cost of developing database systems.

Utilisation of Medicinal Plant Products

To Cultivate the Medicinal Plants

The best way to provide the plant material needed for medicine is to cultivate the plants. This is far better than collecting the plant material from the wild since it does not deplete wild stocks, and in many cases, the declining habitats of native plants can no longer supply the expanding market for medicinal plant products. In the case of rare, endangered or over-exploited plants, cultivation is the only way to provide material without further endangering the survival of those species.

Cultivation also has pharmacological advantages over wild-collecting. Wild-collected plants normally vary in quality and composition, due to environmental and genetic differences. In cultivation, this variation— and the resulting uncertainty of the therapeutic benefit— is much reduced. The plants can be grown in areas of similar climate and soil, they can be irrigated to increase yields and they can be harvested at the right time. Cultivation also greatly reduces the possibility of mis-identification and adulteration.

Many countries have long traditions of cultivating medicinal plants. For example, in some countries, housewives traditionally grow a range of herbs essential for the health care of their families in pots around their houses. Yet in others, most medicinal plants are still collected from the wild; as human populations increase, as forests decline and as areas of remaining wild habitats become locked into parks and reserves, people in some areas can no longer collect the plant materials they need. The need to develop the tradition of cultivating the plants is imperative.

Some mixture of regulations and incentives may be needed to encourage cultivation, especially in parts of the tropics where many medicinal plants are slow-growing trees.

The Ministries of Agriculture and Health should Co-ordinate a programme to establish Nurseries where the Medicinal plants are cultivated. The WHO Traditional Medicine Programme has outlined two basic strategies for the production and use of medicinal plants:

(a) the agro-industrial production and use of medicinal plants of standardised pharmacologically active constituents; and

(b) the distribution of seeds, seedlings and/or saplings to individuals and communities for cultivation in home gardens.

For example, plants with proven antimalarial properties could be distributed to communities during the malaria transmission season. The programme should involve horticultural experts, for example from botanic gardens, to provide expertise on how to grow the plants and to train local staff.

Botanic gardens and Horticultural/agricultural Institutes should improve the agronomy of cultivated Medicinal plants and bring into cultivation those species needed in Medicine that have not been cultivated before.

The range, quality and effectiveness of cultivation techniques varies widely and there is a need for much greater application of modern agronomic methods. There are many ways in which agronomists can improve the ways in which a plant is cultivated. For instance, they can devise better ways of propagation, including novel techniques like tissue culture; they can find the optimum combinations of factors such as light, temperature, mineral and water supply; and they can choose the best time for harvest, so as to maximise the active principle.

Plant Breeding

Plant breeding has long been used in agriculture to produce improved forms of the plants (called cultivars), adapted to cultivation. But so far it has only been applied to very few medicinal plants. Plant breeding can not only improve the agronomic traits of a medicinal plant, such as by optimising the architecture of the plant and improving resistance to pests and diseases, but also improve the phytochemical characteristics of the plant, notably to develop a high and uniform concentration of the active constituents.

Most medicinal plants are at present grown as unimproved wild plants, and so tend to be very variable, even in cultivation. Plant breeding can reduce this variability, and so help standardise the medicinal product. Also, plant breeding can adapt the plants for cultivation in different soils and climates, so that the plants can then be grown under a wide range of conditions.

Effective plant breeding requires a wide range of genetic variation in the plant as starting material. This may either come from wild stocks or from ex situ conservation Botanic gardens in particular have an important role in maintaining wide gene-pools of medicinal plants for plant breeding.

However, the costly and laborious techniques of modern plant breeding are only likely to be used in the cases of plants grown for high-volume crops needed for modern medicine. The investment needed is likely to be too high for many of the plants used in traditional systems of medicine. In this case, a national programme for plant breeding should take their needs into consideration.

The use of weed-killers and pesticides may represent a risk to the environment, to those growing the plants and to those treated with the resulting drug. Therefore, the use of chemicals (if used at all) should be reduced to an absolute minimum, and where chemicals are used, necessary regulations should be in place and rigidly followed.

Provision for training and information is vital. Training may be the best approach for those working in medicinal plant nurseries. Information leaflets and other forms of extension may be more suitable to help members of the public grow their basic home pharmacies themselves.

Sustainability of Wild

Sustainability should be the guiding principle for the ways in which people use nature and natural products. True development can only be sustainable development. If collecting a medicinal plant reduces the wild population, continuing to do so will inevitably impair the rights of future generations.

In the past, many societies had complex systems of rules, often unwritten, to control the exploitation of the natural world. As these traditions have broken down, a legal framework is needed to replace them. The basis for that framework, much simplified, is presented below. So far relatively few countries have laws to protect plants, and those that do are mostly in developed regions like Europe and North America. Yet a legal framework is particularly needed for plants, which are at a legal disadvantage over animals. As private property, plants can be destroyed by the owners of the land on which they grow; yet as free produce of nature, in many countries plants can be collected by anyone almost anywhere.

The ideal system is one where permits are required for any commercial collection of any plant. Otherwise, if a new use is discovered for a plant or if there is a sudden demand for a well-known plant, wild populations may be decimated before the Government can bring in specific controls on collecting it.

If this is not practical, regulations can require permits for the commercial collection of certain named plants that are believed to be in danger of depletion. For example, India bans the collection of Rauwolfia species, except under a permit. The advantage of the permit system, as opposed to blanket bans, is that it gives great flexibility to the issuing authority. If a plant is in steep decline in one area, permits can be suspended there for a year or so, but if a plant is on the increase, the volume of material licensed for collection can be increased. This way, the issuing authority can effectively manage the wild populations.

The regulations should take account of the part of the plant used, and of the capacity of the plant to regenerate afterwards. Collecting fruits, flowers and leaves generally has little impact, but collecting bark, roots and even whole plants can be much more harmful. Regulations should also take into consideration harvesting practices.

The regulations should also respect the ethical, legal and social interests of all those concerned, in particular those living in the area where the plants are native. Regulations can be boosted by public information campaigns, to convince traders and users of the need to conserve wild stocks. Wherever possible, any collecting should be done by trained people, adopting good management and a rotation system, and suitably supervised.

The only exceptions should be to allow the collection of small amounts of the plant as propagating material. This should be done by qualified staff, under permit, and should not further endanger the wild plants. The legislation should ban not only the collection but also the possession and trade of those plants if collected from the wild. This is necessary for enforcement, as it is very hard to catch an offender in the act of collecting a protected plant, but much easier to catch him or her afterwards in possession of the plant.

The legislation should include a prohibition over land-owners collecting material of the plants concerned on their land. When the collection of a plant used by traditional practitioners is banned, measures should be taken to provide them with alternative species that contain products of similar effect. The practitioners should also be involved in the process of conserving the threatened plants concerned.

Trade in medicinal plants should be fully controlled, but in most countries it may only be practical to control trade in and out of the country, making use of customs facilities. Long-established phytosanitary regulations affect the movement of all plants across international frontiers. The Convention on International Trade in Endangered Species (CITES), now with 115 member nations, controls the movement of a large number of plants listed on Appendices, mainly of species threatened by trade and relatives that are difficult to distinguish from them.

If a government wishes to regulate trade in a species not on CITES, the best means may be by requiring export permits. By cultivating the medicinal plants they need, governments may be able to replace imports of medicinal plants with home-grown stocks, and so save foreign exchange, ensure quality and guarantee continuity of supply.

Different Techniques for Harvesting, Storage and Production

The wide array of medicinal plants requires many different techniques for harvesting, storage and production, yet rarely are these researched and documented. For example, the season at which the drug is collected and the age of the plant may affect the amount and nature of the active constituents. The processes involved include selection (of wild plants), harvesting (or collection from the wild), cleaning, extraction of the drug, processing and storage.

Ministries of Agriculture, Health and Trade should develop and co-ordinate a programme to improve the techniques for harvesting and storing medicinal plants and preparing their products. This may best be done as a collaborative effort, involving pharmacological institutes, horticultural organisations and representatives of the users. It should be remembered, too, that local people often have very good knowledge on how to harvest, store and process medicinal plants.

Safety should be the over-riding criterion for use of medicinal plants. To ensure safety, the Ministry of Health should regulate the preparation of drugs and other remedies from medicinal plants used in their country.

Standard reference material of medicinal plants and drugs derived from them is required. In addition to the need for descriptive monographs on plant materials, there is also occasionally a need for reference substances. Until they can be made available from a central point, WHO is assisting in identifying national laboratories which could supply samples of natural substances to be used for reference purposes.

As part of WHO's efforts to establish the safety of medicinal plants, two WHO/DANIDA interregional workshops were held on appropriate methodologies for the selection and use of traditional remedies in national primary health care programmes. These workshops addressed the problems of safety and efficacy involved in the use of traditional remedies, including related issues of standards, stability and dosage formulation. WHO have prepared Guidelines for the assessment of herbal medicines, setting out the procedures for assessment of quality, safety, and efficacy and intended use.

Conservation of Medicinal Plants

The vegetation of the world is being changed or destroyed at an alarming rate. The tropical moist forests, home to about half of the world's plants, are in particular danger, declining at an estimated 16.8 million ha/annum according to UNEP/FAO. Combined with exploitation, this is putting many medicinal plants in grave risk of genetic erosion and even extinction.

The best means of conservation is to ensure that the populations of species of plants and animals continue to grow and evolve in the wild— in their natural habitats. Such *in situ* conservation is achieved both by setting aside areas as nature reserves and national parks (collectively termed "Protected Areas") and by ensuring that as many wild species as possible can continue to survive in managed habitats, such as farms and plantation forests.

Managing large areas of natural vegetation is a complex process and goes far beyond simply removing external threats. The experience of protected area managers over many years has shown that the most important step is the preparation of a management plan, which should define the objectives of the area and outline how they can be met.

The Parks Department should prepare a policy at national level on the conservation and utilisation of medicinal plants in protected areas

The policy should include:

— Identifying which of the protected areas are most important for medicinal plants;

— Targets and techniques for recording and monitoring medicinal plants in protected areas;

— Techniques and procedures for collection of medicinal plants within protected areas;

— A legal mechanism to ensure that benefits reach local people;

— Training of park managers about medicinal plants, including their uses;

— Public education about medicinal plants in protected areas.

Protected areas that are important sites for wild medicinal plants are good sites for continuing research programmes on genetic improvement of medicinal plants. Linkage with local universities and botanic gardens can often be extremely useful.

The policy should be at national level, but should be implemented by individual park managers. It should be prepared in consultation with conservation and development groups, and other relevant sectors, in particular the Ministry of Health.

In general, the objective of conserving medicinal plants may be compatible with the other objectives of protected areas, such as watershed protection and saving biodiversity. But this will not always be so: in some cases, a management objective, such as promoting tourism or encouraging climax vegetation, may conflict with the need to conserve a medicinal plant. In such cases, certain areas would need to be zoned and conflicting uses controlled.

The Parks Department should assess the extent to which the protected areas system covers the medicinal plants of the country. It should then create new protected areas and extend existing ones to ensure that all the medicinal plants of the country are conserved. The starting point for this process is an up-to-date map of the vegetation cover of the country. Images from satellites provide a way to keep this information up-to-date.

Once the habitat of a species has been identified, botanists can predict where that species might occur, in addition to those places where it has already been found. This enables maps to be prepared highlighting the key areas for medicinal plants. Species which only occur in areas of natural vegetation— as opposed to species in disturbed habitats (such as *Catharanthus roseus*, Rosy Periwinckle, which, outside its natural habitat, is a roadside weed)— should be given priority. Of these species, the endemics— plants confined to the country— should receive particular attention.

Park planners can then superimpose maps of:

— species distributions;

— surviving vegetation; and

— existing protected areas.

This will show the extent to which the protected area system covers the medicinal plants of the country, and where the principal gaps in the system occur. Although this can be done manually, using maps drawn on transparent plastic, it is better done nowadays on computers, using the technology of Geographical Information Systems (GIS).

The Parks Department should devise economic and social incentives for maintaining natural habitats and wild species.

Modern conservation planners try to reduce the loss of living resources by showing how conserving biodiversity helps to safeguard the benefits that people derive from wild species. Medicinal plants are only one of these benefits and should be treated together with other benefits, such as protecting relatives of food crops and maintaining supplies of fresh water.

Park managers should ensure that the some of the benefits of conserving medicinal plants accrue to local people. In the past, protected areas tended to exclude local people, in some cases even from their traditional lands and practices, but this is now seen as a mistake. Large protected areas will only survive if they have the support of the people who live nearby. Thus conservation planners try to ensure not only that local people do not lose out from the establishment of a protected area but also that they actually benefit.

Park managers should ensure that the conservation and exploitation of medicinal plants are incorporated into site management plans.

In some areas, it may be appropriate to allow local people to collect limited amounts of medicinal plants in protected areas for their own use. This can improve public relations for the park, and may help to discourage illegal and damaging activities. But it should only be permitted to levels that are sustainable, and a percentage of the profits earned from the medicinal plants should be returned to improving management of the protected areas. Where a commercial company makes use of plants from the protected area, arrangements should be made to ensure that the industry contributes to local needs— of both park and people.

In other areas, a policy of absolute protection, except for removal of small amounts of propagating material, may be appropriate. The pharmacologically active ingredients in medicinal plants may vary within the same species, depending on factors such as soil chemistry, type of vegetation, and presence of insect predators. For this reason, to conserve the full genetic diversity of the species, a range of wild populations of each medicinal plant should be maintained, even when the main source of supply is cultivation.

Park departments can also help encourage the process of cultivating the medicinal plants. They can provide seeds and saplings for local people, and can even set up small medicinal plant nurseries in suitable areas.

Species that are heavily depleted by over-collection should be re-introduced into areas where they once grew wild.

In appropriate cases, species which have become extinct in nature may be re-introduced either to their original localities or if that no longer exists into a similar habitat. Guidelines for such re-introductions are being prepared by Botanic Gardens Conservation International and IUCN's Species Survival Commission.

Conserve Populations of Medicinal Plant Species Ex situ

Ideally all medicinal plant species should be conserved as evolving populations in nature. However, these species should also be conserved ex situ (i.e. outside their habitat) as well. The primary purpose of this is as an insurance policy. But it also has the advantage that it is usually easier to supply plant material for propagation, for re-introduction, for agronomic improvement, for research and for education purposes from ex situ collections than from in situ reserves.

The disadvantages of ex situ conservation are that the sample of the species conserved ex situ may represent a narrower range of genetic variation than that which occurs in the wild. Species conserved ex situ can also suffer genetic erosion and depend on continued human care. For this reason, ex situ conservation must not replace, but should complement, in situ conservation. Most important of all, ex situ conservation should not be used as a reason for failing to safeguard representative samples of the medicinal plants and their habitats in nature.

Priority for ex situ conservation should be given to species whose habitats may have been destroyed or cannot be safeguarded. It should also be used to bulk up populations of depleted or even locally extinct plants for restocking in nature. In some countries it may be appropriate to conserve all medicinal plants ex situ, in others, where for example some medicinal plants are common weedy species, this may not be necessary. When collecting the plant material for ex situ conservation, care should be taken not to put the survival of the wild population at risk.

In most countries, botanic gardens are the organisations best suited to carrying out the ex situ conservation of medicinal plants. The origins of botanic gardens, in Europe in the 16th Century, were as training centres on medicinal plants, and today many botanic gardens, especially in Asia, have important living collections of medicinal plants. Botanic gardens are also important centres for research and monitoring of wild plant populations, and have a vital educational role as the shop window of botany to the outside world.

Seed banks are the best form of ex situ conservation for plants. In a seed bank very large numbers of species and individuals can be stored in a small space, but seed banks are vulnerable to loss of electricity supplies, require careful monitoring and do need time-consuming regeneration.

Running a seed bank is a specialised and full-time job, requiring meticulous record-keeping, seedtesting and "growing-out". Little is known of the seed biology of most medicinal plants; in many cases research will be needed to adapt standard techniques to the needs of individual species, and therefore, most important of all, seed banks should include a research facility to study seed biology and population genetics.

One problem is that with existing techniques the seeds of an estimated 50,000 plant species— 20% of the world's total— cannot be stored without loss of viability. These species are termed recalcitrant and for them other techniques are needed. Some other species may not produce seed at all.

Botanic gardens should set up alternative means of ex situ conservation for those species which cannot be stored in seed banks.

For plants where seed banking is not possible, if they are long-lived species like trees and shrubs, the best means of ex situ conservation is the Field Genebank. This is simply a plot of land in which the plants are grown in rows and carefully labelled. The disadvantages are that Field Genebanks take up a lot of space, cannot conserve as much variation as in seed banks, and are vulnerable to disease epidemics.

In fact many samples of medicinal plants are already grown in botanic gardens. In some cases, these collections are in the public area of the garden, but in many cases the material is in cold frames and greenhouses. This provides a good temporary measure of ex situ conservation as well as providing material for various purposes. Conventional cultivation in flower beds is ideal for demonstrating the interest and importance of medicinal plants to the public, but is of little value for genetic conservation because of the few specimens of each species usually held.

Effective techniques of ex situ conservation other than seed banks or Field Genebanks are still in their infancy and need further research before they can be recommended for general use. The main approach under study is cryopreservation— that is the storage of tissue and cell culture at low temperatures, in some cases as low as—196°C.

Communication and Cooperation of Medicinal Plants

Building public support for the conservation of medicinal plants is essential and worthwhile for several reasons. To create such support first requires public understanding.

In part due to encouragement from WHO, an increasing number of governments and other agencies are fully aware of the importance of medicinal plants in providing primary health care to their people. Many experts and practitioners now have an in-depth understanding of the issues involved. But without a solid foundation of understanding by the public, essential research and management programmes may not be fully secure at times of political change or financial stringency.

Unless the public, who are the ultimate beneficiaries of medicinal plant resources, fully appreciate the work that needs to go into the conservation of those resources, vital political backing may prove to be fragile. To this end, ministries, agencies and others will find it worthwhile to conduct well-targetted public information and education programmes.

Wherever possible, scientific experts and administrators should obtain the assistance of communication professionals in drawing up a strategy for communicating the importance of the conservation of medicinal plants.

The strategy should cover communication to the public and to health practitioners at a local level— in effect external communication. It should also be targetted to the relevant policy-makers, administrators and scientists both inside and outside government— in effect internal communication. Many well drawn up communication strategies fail because the organisation itself is not seen to "practice what it preaches" and this is often the result of a failure of internal communication.

Within ministries and agencies, professional staff such as press or information officers will normally be on hand, but their time is frequently taken up with the day-to-day priorities of the organisation, and there is a risk that if no distinct, self-contained and comprehensive strategy is drawn up on medicinal plants, the issue and message may get swamped or diluted by other messages, which are justifiably seen as important. Indeed, given the difficulties of delivering effective health care in any country, there is a tendency for communicators in an organisation having to deal with the urgent at the expense of the important.

For these reasons it is important to draw up a written communication strategy and to identify the means of implementing it, rather than simply rely on the usual channels to deliver the right message automatically.

In-house communications staff must be involved in outlining the process, and are essential in implementing it. But there is often a good case for bringing in additional help from outside the organisation. Not only would such advisers bring a fresh eye to the problem, but they may be able to devise a strategy that functions well across several agencies or institutions. They will also be able to devote their time to the task in a single-minded way that in-house staff can rarely do. In-house staff should still be involved at each stage, for example by drawing up the brief, selecting the relevant experts, providing them with information, and helping organise the "selling-in" of their work to as many staff as possible within the organisation.

Given the support of conservation groups for this "good cause", local advertising agencies or other communication companies may second staff or undertake work free or at a reduced rate. Even if this is not so, asking companies for assistance is a good way to make the communication community aware that medicinal plant conservation is a subject of importance.

There is no single blueprint for a communications strategy but a few principles can be identified:

— *Effect desired* Identify what the audience is expected to do as a result of receiving the message.

— *Core message* Define the message in terms which will persuade the particular audience to take the action desired. Deliver only one principal message in each piece of communication.

— *Targetting* Draw up separate communication activities and materials for different target audiences.

— *Planning* Produce a written plan for communication activities identifying who will undertake which actions in which weeks and months.

— *Research and Monitoring* If possible assess the existing perceptions and levels of awareness of the subject among people in the target groups before the campaign starts, and repeat after the campaign. In this way it is possible to measure the results of the work with some objectivity.

To be effective, communication strategies are most effective when they are targetted at specific groups. In the case of conservation of medicinal plants, there are many groups involved. First and foremost are the users— health professionals, including traditional health practitioners. Then there are the relevant professions, including agronomists, horticulturists, pharmacologists, park managers and plant breeders. In Government, they include policy-makers and legislators. In industry, they include business leaders, in particular from food and phaımaceutical companies. In the media, they include newspaper editors, health and environment correspondents, and TV producers. They also include community and religious leaders as well as staff in conservation organisations.

Communication strategies work best when the aim is for the target audience to take a particular action as a result of the message. To do this, the first step is to raise the awareness of the group concerned on the conservation of medicinal plants as an issue. Awareness itself rarely achieves practical results— it is not, for example, a sufficient condition to cause a desired result— but it is an essential precursor to achieving practical results.

Some possible changes that the campaign might seek are listed below:

— Improved training in the professions. In this case, as with any dealings with professionals, it is best to use people from that profession in selling ideas or information to that profession.

— Improved practices in the professions. In general the professional associations are helpful in selling-in new ideas if they can be persuaded of the merits of the case. The editors of professional magazines are also often particularly helpful. Information targetted at a profession from outside, however, may be resented.

— Incorporation of material into the formal education curricula. It is normally advisable to use existing machinery through established channels, as most educational systems are planned on the longer term. This also presents an opportunity to seek leverage by using existing educational media and budgets to carry the message, and to persuade educational publishers, broadcasters, etc., to do so, rather than to produce new materials. Mass production of formal educational materials is not usually cost-effective in countries with a reasonable educational infrastructure.

— Incorporation of information into informal education and outreach. This is an enormous area of potential activity and will vary according to national and local circumstances. In general, however, it is worthwhile for ministries and agencies to work in partnership with nongovernmental organisations, particularly when these have at least some professional staff and a capacity to manage projects. Funding projects such as an education officer on medicinal plants will often unlock large amounts of pro bono voluntary resources otherwise unavailable to government.

— Encouraging villagers to grow the medicinal plants they need rather than uproot them from nature. This may also be a worthwhile task for schools, youth groups, religious groups, universities and businesses. It can be stimulated by government and other authorities giving them seeds, saplings and other propagating material free of charge.

Examples of approaches

— Medicinal Plant Gardens. A collection of medicinal plants within a public parks or botanic gardens, or even in the grounds of hospitals, clinics, forestry stations, hotels or local authorities, can form an intriguing, accessible and popular feature. The garden can be laid out by reference to parts of the body showing the ailments the plants may be used to treat.

— Guided visits and open-days at research facilities and botanic institutes. These events are equally popular with the public and with professionals, although events for these audiences should probably be separated. When, for example, health professionals make such visits, it is important to have their "peers" on hand to discuss technical matters with them. The public may be most interested in events where they can get "hands-on" experience of the therapeutic use of medicinal plants, especially for self-medication.

— Lectures and formal courses. Lectures and talks may be arranged at the place of work for particular target groups (for example, civil servants in a ministry or doctors in a hospital). In this way, a useful question-and-answer session may develop, which is not possible with even very expensive materials such as exhibitions. Also, a public institution may arrange a series of lectures to the public, and publicise it accordingly.

— Educational Campaigns. The attention of the press and public may often be best aroused by a time-limited "campaign" of public awareness, rather than essential but routine work. While ministries and agencies can play a direct role in such campaigns, it is often better to fund a non-governmental organisation to undertake the main activities. For example, the annual treeplanting day could be used to promote the cultivation of medicinal plants in home gardens.

— Protected areas. Many national parks are under-used and few people realise that the plants used in popular remedies can actually be seen in their wild habitat. Efforts should therefore be made to attract people to such areas and to see the plants growing in their natural environment. Where the plants are sustainably harvested, information materials should explain that this does not harm the plant population or the purposes for which the park was established.

10

Marketing of Medicinal Herbs

Plants have been used since ancient times to heal and cure diseases and to improve health and wellbeing. Despite ancient nature of the tradition, medicinal plants still form the basis of traditional or indigenous health systems and are reported by the World Health Organisation (WHO) to still be used by the majority of the populations in most developing countries.

Medicinal and aromatic plants (MAPs) play a significant role in meeting the demands of the traditional medicine markets which are found both domestically in the producing and in overseas markets. Traditional medicine, for example, which is related to traditional Chinese medicine (TCM), Indian ayurveda and Arabic unani medicine and to various forms of indigenous medicine, as well as the complementary or alternative medicine utilised in industrialised countries, is achieving growing credibility in many parts of the world. The percentage of people using traditional medicines decreases in developed countries: 40-50 percent in Germany, 42 percent in the USA, 48 percent in Australia and 49 percent in France.

The last three decades have seen substantial growth in herb and herbal product markets across the world. Rapidly rising exports of medicinal plants during the past decade attests to worldwide interest in these products as well as in traditional health systems. According to the Secretariat of the Convention on Biological Diversity, global sales of herbal products totalled an estimated US$60 000 million in 2002.

At present, 80 percent of the population in developing countries rely largely on plant-based drugs for their health care needs, and the WHO has estimated that in coming decades a similar percentage of the world population may well rely on plant-based medicines. Thirty percent of the drugs sold worldwide contain compounds derived from plant material.

As a result of the expanding interest in medicinal and aromatic plants, new income-generating opportunities are opening up for rural populations. With many of the MAPs gathered from the wild, the collection and sale of MAPs is providing a complementary source of cash for many extremely poor rural households. However, despite the fact that the products collected can have very high value in the final products, the collectors typically receive only a small share of the final value, either because they are unaware of the real value, are unable to market it in the form wanted by buyers or are unable to market to these buyers.

PRODUCTION

There are two sources of supply of medicinal plants:

i) material collected from the wild; and

ii) cultivated material.

Material from Wild Sources

Wild harvesting is the collection of plant material from wild sources. This material can take many forms, such as the bark, leaves, fruits, herbs, flowers, wood or roots. It may be collected from many locations, including open pasture, waste agricultural land, gardens, the roadside or forest land. In some cases the plants may be "weeds" found in agricultural or waste land; in others they may be plants or parts of plants found in horticultural areas or in forest land.

The bulk of the material traded (both domestically and internationally) is still wild harvested and only a very small number of species are cultivated. It is difficult to provide accurate global data on the volume of wild harvested medicinal plants as it is very difficult to distinguish between wild and cultivated material. Although data can be indicated for some specific cases, there is very little actual global data on the volume of wild harvested medicinal plants.

Herbal raw material is often either collected by wage labourers (often from outside the state) or farmers with small landholdings. Cultivation of herbal raw material is rare: in Bangladesh, for example, more than 90 percent of the collection of medicinal plants is from the wild. Illegal and unscientific collection is common. Although the major part of wild harvested material is sourced from developing countries, a surprisingly high amount is also gathered in developed countries.

Of major concern is the fact that a significant part of wild harvested material is now traded commercially. As the prices paid to the gatherers tend to be very low, commercial plant gatherers of "mine" the natural resources rather than manage them, as their main objective is to generate an income. Although there are many common species that can be harvested sustainably and with little impact on their survival, an increasing number is not in this category. Of particular concern for the sustainability of the wild resource is the fact that many of the materials are the roots of plants, which are the most difficult plant parts to harvest sustainably.

A critical factor in wild harvesting is the availability of cheap labour to undertake the very labour intensive work of gathering. Because in many cases income from such sources represents the only form of paid employment for inhabitants of remote rural areas, there is a ready availability of workers. Further, contractors who employ the collectors often act as middlemen and traders as well. Collectors are often dependent on contractors as they are poor and often owe money to the contractors.

Most countries have few or no regulations which control the collection of material from the wild. India, Bulgaria and Nepal are notable exceptions. India has banned the export of several wild species in their raw material form, although the export of finished products containing the material is allowed. A major part of the high range Himalayan plants are wild harvested and many of these are close to extinction from over-harvesting or unskilful harvesting: *Nardostachys jatamansi*, Aconitum spp., for example.

Cultivated Material

Cultivated material is more suitable for large scale uses, such as the production of drugs by pharmaceutical companies, which require standardised products of guaranteed or known content and quality. These quality requirements are becoming increasingly important as drug regulations become more stringent in many countries.

Argentina, China, Hungary, India, Poland and Spain are examples of countries that cultivate some materials on a large scale. Requirements of successful commercial cultivation operations are to produce high quality drugs using low input cultivation methods while recognising that the material has to compete on a highly competitive international market.

Given the higher cost of cultivated material, cultivation is often done under contract. In the majority of cases, companies tend to cultivate only those plant species which they use in large quantities or in the production of derivatives and isolates, for which standardisation is essential and quality is critical. Some grower cooperatives or collaborative ventures have been set up to enable growers in some countries to achieve greater negotiating power and achieve higher prices. These have tended to be in developed countries, such as the Netherlands, and in Eastern European countries which export significant quantities to the rest of Europe and to North America.

Globally, the areas cultivated are limited because cultivated material bears higher production costs, must have secure land ownership or access and requires more sophisticated (and costly) management expertise. Costs must be carried for long periods— in many cases over ten years. The low prices of wild harvested material make the return to cultivation low in many cases.

DISTRIBUTION

The supply chain is often very long with as many as six or seven marketing stages involving primary collectors and producers, local contractors, regional wholesale markets, large wholesale markets and specialised suppliers.

The long supply chain contributes to the low prices primary collectors and farmers receive for their products. As collection is still more common than cultivation, huge differences in the quality of raw materials occur. The differences concern the amount of active ingredients based on where the plants were grown, what parts of the plants are being used, how the plants were harvested and how they were stored. Raw material is often also adulterated as collection from the wild cannot guarantee the uniformity of raw material.

Industry buys from suppliers and wholesalers rather than direct from smallholders because of the substantial quantities and broad range of raw material that is needed. This makes product traceability nearly impossible. Currently, contract farming and buy-back arrangements provide the only practical alternatives for exporters whose customers require traceability.

Improving Practices

Cultivation of medicinal plants faces a number of problems, partly due to the typically small scale of operation. These include the following:

- The majority of farmers have small land holdings;
- Shortage of labour in rural high altitude areas;
- Long period between crop growing and harvesting;
- Bureaucratic difficulties in obtaining permits for cultivating restricted species;
- Lack of technology and difficulties in cultivating medicinal plants (particularly in high altitude areas);
- Even if cultivation technologies are developed, problems with packaging, storage, transportation and quality control persist and are neglected;
- Experiences as well as the needs of farmers are often not included in the research activities of the laboratories;

— The link between research institutes and industry is weak;

— The lack of planting material and the poor quality of planting material; and

— Prices are too low to make cultivation attractive.

These features contribute to the limitations faced by many developing countries, and some are equally applicable to wild harvested plants as to those cultivated.

Market Conditions

Medicinal and aromatic plants are traded both as commodities and, in some cases, as processed final products. Demand for a wide variety of species is increasing as these markets expand and new end-uses are developed.

Each of these groups can contain a wide range of products. For example, herbal products can be used in the following product categories:

— Herbal medicinal products.

  — conventional,

  — traditional,

— Food supplements/dietary supplements.

— Foodstuffs.

— Cosmetics.

Examples where the highest use of traditional medicines are practiced include:

— 100 percent in the United Arab Emirates,

— 100 percent in China,

— 70 percent in India,

— 70 percent in Pakistan,

— 70-80 percent in Africa.

Pharmaceutical companies use medicinal plants:

— For the isolation of single purified drugs;

— In advanced extract form where the extract is highly standardised in terms of the active constituents it contains. In many cases, these are in admixtures with other ingredients;

— As starting material for the production of other semi-synthetic pharmacologically active substances. For example, plant saponins can be extracted and altered chemically to produce sapogenins required to manufacture steroids.

A significant percentage of medicinal plant material is used to make plant extracts. This is carried out either by the end product manufacturers or by extract companies. In addition to the market for medicinal plants, there is an expanding market in developed countries for botanicalbased products, such as health foods and supplements, herbal drinks and various health and personal care products. The market for herbal products throughout the world is currently worth around an estimated US$60 billion per year with a growth rate of 7 percent.

Market Features in Preferred Countries

China

The market in China is large and shared between public and private ownership. Thirteen of the top companies which produce TCMs are listed publicly on the domestic stock exchange. Fourteen are state-owned.

Between 1990 and 2001 the total area of medicinal crops increased from 153 000 to 827 000 hectares. This underlines the Chinese leadership in the sector. China's planted area is some 20 percent greater than the USA, the second largest supplier.

It should be underlined that TCM tends to use the roots of plants, which are the most difficult plant parts to harvest sustainably. However, GACP for production of herbal materials in China can be traced back to more than two thousand years. The first work on Chinese herbology mentioned that practitioners should pay more attention to the influences of such thins as the environment and location, soil and water, time and season, process and store and so on for cultivation and collection of herbs. Over time, TCM has developed a theory of geo-herbalism which has been used in the evaluation of efficacy, quality and safety, making TCM well known in the world.

The GAP system was introduced into China from Europe in 1999 and has been adopted officially on 1 June 2003. GAP plantations of more than 30 medical species, such as peony, ginseng, liquorice and agelica, are now distributed throughout China. Some manufacturers have also been involved in GAP. However, GAP is seen as a heavy burden for the production of traditional medicines in developing countries as it has increased costs and reduced yields against already low returns to producers. In developed countries, GAP is relatively easier to implement but its introduction should result in improvement and standardisation of TCM.

India

India has a long-standing knowledge and understanding of using medicinal plants in its codified medical systems, as well as through its highly diverse folk traditions. The knowledge of plant identification, methods of collection and processing, biological activities and uses was transmitted both orally and in written form by traditional medical cultures in India.

Medicinal plant trade in India is substantial with total turnover of Rs. 2 300 cores of ayurvedic and herbal products, while major over-the-counter products contribute around Rs. 1 200 cores. Despite its substantial domestic trade and its long experience with herbal medicines, India has not been able to capitalise on this by promoting use in developed country markets. To achieve improvement in this area, India must identify products which may be relevant to diseases found in the developed world and for which no medicine or only palliative therapy is currently available. This would enable rapid access of these herbal medicines into developed country markets.

There are major challenges in tapping the substantial potential for utilising medicinal, aromatic and natural dyes plants (MADPs) nationally in India, as well as in export markets. At the forefront of these problems is ensuring consistent and acceptable quality. Traditionally, because the usage of plants was a part of a local community's culture and health practices quality was more manageable. However, this tradition is being rapidly eroded as control of medicinal development and usage moves to the industry.

Quality in the case of medicinal plants includes identity, purity and safety, as well as efficacy. However, there are many challenges involved in quality standardisation of MADPs.

These include the following:

— Unlike allopathic drugs that have a single or a set number of compounds that can be easily standardised, plants have a plethora of phytoconstituents that contribute to their bioactivity;

- Bioactivity also varies according to time (day, season, constellar position) and to region (arid, marshy);
- There are differences in the bioactivity depending on the way the plant has been collected, processed and stored.

Phytoconstituents that attribute specific medicinal properties to a plant are not present in uniform quality or quantity under different edaphic conditions. The medicinal and toxic property of a plant may also vary according to the kind of processing used. Traditional, ancient Indian systems of medicine recognised this and had specific recommendations for the collection of medicinal plants and the preparation of formulations in order to obtain the best efficacy.

The current relevance of the recommendations for selected medicinal plants needs to be determined using modern analytical tools such as chromatography and bioassay. These techniques would indicate, respectively, the differences in the chemical profiles and their significance on the bioactivity of plants collected and/or prepared according to traditional advice.

The potential growth in domestic market demand in developing countries is exemplified by the growth projections for India by the National Medicinal Plants Board. These projections indicate a high annual growth rate for many of the species.

Malaysia

In Malaysia there are three sources of medicinal plants used by the industry, namely from the wild (forests), from plantations and through imports. Previous uncontrolled collection of herbal material has resulted in depletion of the wild resources. A study conducted by Azmi & Norini to estimate the demand and supply of medicinal plants from the forests by both collectors and the herbal industry in northern states of peninsular Malaysia estimated that the collectors may collect between 300 to 500 kg of medicinal plants on every collection trip, with an average of 2.67 trips per month.

The herbal industry in Malaysia is facing a shortage of local raw materials to meet the demands of the growing industry, with the result that the industry is heavily dependent on imports for its raw materials and processed herbs. This is highlighted in the trade statistics which show that imports of medicinal and aromatic plants between 1995-1999 exceeded exports by a wide margin. The majority of the imports came from China, India and Indonesia. Singapore and Thailand were the major importers of medicinal and aromatic plants from Malaysia.

The Malaysian herbal industry has grown rapidly and has the potential to become a significant industry in Malaysian agriculture. Botanicals (including their extracts, exudates and essential oils) have found application in a wide range of products that includes food and beverages, pharmaceuticals, herbal/traditional medicines, health care, heal enhancing products, dietary supplements, flavours and fragrances, cosmetics and toiletries and various consumer goods, such as dyes, detergents, biopesticides and other industrial chemicals. They can be classified into four major product groups.

The Malaysian Third National Agricultural Policy (1998-2010) identified the Speciality Natural Products Industry as a "new and future" industry group. This group refers to natural products based on botanicals (herbs, medicinal plants, spices and aromatic plants) and aquatic plants and animals. These have been showing considerable growth in recent years, with the pharmaceutical market growing at about 10 to 15 percent per annum, the health related food product industry growing at 20 percent per annum and the market for cosmetics and toiletries growing at 10 percent per annum.

OECD Countries

In the OECD group, the EU, United States and Japan are the most involved in the medicinal plants trade. In these, imports have always been greater (often substantially) than exports. As shown below EU imports of medicinal plants, excluding intra-EU trade, are higher than those into the US and Japan, while exports (by value) are considerably higher than from the US when intra-EU trade is taken into account.

Almost 95 percent of the total EU imports of medicinal plants (HS 1211) (intra- and extra-EU) is of plants, parts of plants, seeds and fruit, used in perfumery, medicaments or for insecticidal, fungicidal or similar purposes, fresh or dried, cut and uncut, crushed or powdered (HS 121190). The remaining five percent are imports of liquorice roots (HS 121110) and ginger roots (HS 121190). The main suppliers of ginseng roots into the EU are China, the Republic of Korea and the US, whilst the main suppliers of liquorice roots are Turkmenistan, China and Azerbaijan.

EU exports of medicinal plants are mainly intra-EU. The main extra-EU markets are the US, Brazil and Japan, which mainly import liquorice roots fresh or dried from the EU, while Australia, the US and Switzerland are the main importers of ginseng roots.

Herbal products can be marketed in Europe as either medicinal products or food products. Regulatory requirements in the EU are strict and growing (as with those in many developed countries). Claims made regarding medicinal products may only relate to the treatment and the prevention of disease for which the product has been authorised. Claims for foodstuffs may not relate to the treatment or prevention of a disease and only to normal body functions.

An EU-wide regulation on nutrition and health claims is currently under discussion and will strictly control the scientific substantiation of claims for food products. However, it will allow the use of disease risk-reduction claims, if they can be substantiated by scientific background and/or clinical data. Legislation on traditional herbal medicinal products was harmonised in early 2004 by Directive 2004/24/EG. This Directive will be fully applicable once incorporated into the national legislation of the 25 Member States of the EU by 31 October 2005.

Traditional herbal medicinal products are characterised by the fact that their efficacy can neither be demonstrated by clinical studies nor by bibliographical data. Their efficacy may only be apparent from long term traditional use and experience. Long term use is defined as a minimum use for 30 years, of which at least 15 years of safe use must be demonstrated within the Community. If the applicant can only demonstrate less than 15 years within the Community, the country where the registration is filed may submit the application to the European Medicines Agency (EMEA) for decision.

International Trade in Medicinal Plants

Trade in medicinal plants is difficult to estimate accurately because much of the local trade is either unrecorded or poorly classified and because medicinal plants are also used in nonmedicinal end-uses and not reported separately. Domestic trade, in particular, is poorly recorded. Rising global interest in medicinal plants has also created a sustained and largely "underground" trade in plant materials, many of which are being collected in LDCs in an unregulated manner, resulting in indiscriminate harvest of wild varieties and serious damage to biodiversity. It is, therefore, not possible to assess global trade in all medicinal plants. In addition, official trade statistics either do not identify the plants individually or do not separate their medicinal use from other usage.

Medicinal plants are, however, inputs into a wide range of materials that are used in medicinal or health products and modified forms or different products which contain elements of these plants are on-trade. Germany, for example, is one of the main importers of medicinal plants and also a major international producer and exporter of medicines.

Accepting these data limitations, the following information provides a broad picture of the international trade although, for the above reasons, the data shown should only be considered as indicative.

In volume terms both global exports and imports have been increasing, although the total value has been declining, suggesting falling average unit prices. Despite the fact that consumers in developed countries are not using traditional medicines as frequently as those in developing countries, in absolute terms developed and developing countries import similar values of medicinal plants. Developing countries, however, are the dominant exporters.

Exporters and Importers

Markets for herbal medicines in developed countries— especially in Europe and the United States— are highly regulated and very difficult to penetrate, particularly for developing countries and LDCs whose materials have not undergone the stringent tests required by developed country pharmaceutical manufacturers before mass production. Developed countries therefore tend to export unprocessed or slightly processed materials.

In the case of India, around 80 percent is export of raw materials including dried plants, extracts and isolated ingredients. The export of finished medicinal products, mostly homoeopathic and ayurvedic medicines, accounts for the remaining 20 percent. Some developing countries with a long tradition of use of medicinal plants are major exporting countries— China, the Republic of Korea, Chile, India, Brazil and Thailand, for example. Exports are predominantly in raw material form and only to a lesser extent finished products. With their large populations and ancient heritage of traditional herbal-based medicines, China and India are two of the world's largest markets for medicinal plants, though not necessarily the largest traders.

The main importing countries are, by value, China, Hong Kong Special Administrative Region (SAR); the US; Japan and Germany. The latter is the leading importer within Europe because its pharmaceutical companies are major players in the world market. The market for herbal products is very diverse throughout the world, with each region or each country having its own prerequisites for bringing those products on the market.

The classification of the products may also vary widely among the different countries. In one country a herbal substance may be classified as medicine, in another as food. Food and medicine often require different quality approaches, as the quality assurance systems used for food (hazard analysis and critical control point [HACCP] or food GMO, for example) tend to differ from the systems to be used for medicines (pharmaceutical good manufacturing practice [GMP]). Additionally, the heavy metal and residue limits tend to vary between the different product groups. It is therefore important that these variations are accommodated and guaranteed by the country of origin.

In order to meet these types of variations, the basic requirements for gaining entry into developed country markets are evidence of:

— Well documented traditional usage;

— Single-plant medicines;

— Medicinal plants free from pesticides, heavy metals, etc;

— Standardisation based on chemical activity profiles; and

— Safety and stability.

Different Pricing Points

It is very difficult to generalise regarding prices because of the wide range of products and materials included as medicinal plants. This highlights the variability shown in different markets for different materials and product forms, and between different pricing points (for example, fob and cif). It emphasises that any assessment of market potential needs to be done at a product and market level.

At a macro level, it appears that average real prices were falling over the period 1991-2002. As with many agricultural products in developed countries the share received by local producers and gatherers for raw plants is usually low. Since the cost of production in organised cultivation is invariably higher than the prices paid for wild collections, it is difficult to persuade communities to undertake organised cultivation. Research in Nepal has clearly indicated that there is a significant difference between the returns from wild collections and those from cultivation of medicinal plant species in that country. The differences in profitability have been, and remain, a serious disincentive to organised cultivation.

Tariff Rates

Imports into many markets are affected by government import restrictions. The severity of such restrictions can affect the accessibility of some markets to imports of medicinal plants. The most obvious of these are tariffs, but others such as quotas, phytosanitary standards, and packing requirements may also exist and create difficulties for medicinal plant trade.

In general, rates on most materials into developed countries are zero or low. Those on a number of products into developing countries which have a local industry they wish to protect are substantially higher (although there may be exemption to some rates). For example China indicates rates of 22 percent ad valorem on some of its imports; India has a rate of 15 percent (with some exemptions); Turkey 35 percent; and Bangladesh 22.5 percent. This level of tariff makes exporting to these markets extremely difficult. On the other hand tariffs into the EU, USA and Japan are zero or very low.

Bibliography

Baer, Mary E., "Pot-pourri album", *The herbarist,* 1954.

Baricevic, Dea and Tomaz Bartol., "The biological/pharmacological activity of the Origanum genus", *In Oregano: the genera Origanum and Lippia,* Edited by Spiridon E. Kintzios, London: Taylor & Francis, 2002.

Becker, Jim, "Herbs are for butterflies, too", *The herbarist,* 2000.

Belsinger, Susan and Carolyn Dille., "Cooking with marjoram and oregano", *The herb companion,* 1992.

________________ , Susan and Tina Marie Wilcox., "Our favorite oregano in the kitchen", *The herbarist,* 2004.

Fox, Helen M., "Herbs in knots and laces", *The herbarist,* 1953.

Gardner, JoAnn, "Holy land herbs revisited", *The herbarist,* 1989.

Harpaz, S., et al., "Effects of herbal essential oils used to extend the shelf life of freshwater-reared Asian sea bass fish (*Lates calcarifer*)", *Journal of food protection,* 2003.

Jones, Dorothy Bovee, "The great dittany of Crete – then and now", *The herbarist,* 1973.

Kelaidis, Panayoti and Gwen Kelaidis., "Herbs of the rocks: the Mediterranean garden", *The herbarist,* 1988.

Kintzios, Spiridon E. "Profile of the multifaceted prince of the herbs", *In Oregano: the genera Origanum and Lippia,* Edited by Spiridon E. Kintzios, London: Taylor & Francis, 2002.

Lord, Priscilla Sawyer., "The Herb Society of America – a kaleidoscopic view of the first 50 years", *The herbarist,* 1983.

Makinen, Seija Marjatta and Kirsti Kaarina Paakkonen., "Processing, effects and use of oregano and marjoram in foodstuffs and in food preparation", *In Oregano: the genera Origanum and Lippia,* Edited by Spiridon E. Kintzios, London: Taylor & Francis, 2002.

Makri, Olga., "Cultivation of oregano", *In Oregano: the genera Origanum and Lippia,* Edited by Spiridon E. Kintzios, London: Taylor & Francis, 2002.

Rollins, Elizabeth., "Origanums: beauty and fragrance in the summertime", *The herbarist,* 1995.

Skoula, Melpomeni and Jeffrey B. Harborne., "The taxonomy and chemistry of Origanum", *In Oregano: the genera Origanum and Lippia,* Edited by Spiridon E. Kintzios, London: Taylor & Francis, 2002.

Small, Ernest, Brenda Brookes and Marcel Jomphe, "A bee that loves herbs", *The herbarist,* 1995.

Talbert, Rexford., "Amaracus", *The herbarist,* 2004.

_______________ , Rexford, "It's all in the calyx", *The herbarist,* 2004.

Tucker, Arthur O., "Will the real oregano please stand up?", *The herb companion,* 1992.

Webster, Helen Noyes, "Notes on the marjorams", *The herbarist,* 1948.

Wilcox, Tina Marie and Susan Belsinger, "Don't let pruning get you down", *The herb companion,* 2004.

_______________ , Tina Marie and Susan Belsinger, "Mediterranean companions", *The herb companion,* 2004.

Zava, D. T., Dollbaum C. M. and Blen.M. "Estrogen and progestine bioactivity of foods, herbs, and spices", *Proceedings Society Experimental Biological Medicine,* 1998.

Index